J. F. Fitch 1

2 How To Be A Math Genius

TABLE OF CONTENTS

INTRODUCTORY TIPS

Motivation: This book is mainly for people who already understand the basic concepts of math and wish to be able to do math in their head. If math isn't your best subject, you may only understand some of the tricks in this book. Do not get discouraged if you can't understand all of them. You can skip the ones you don't understand.

Pencil and Paper: Before being able to do a method in your head, you'll have to first be able to understand it. When practicing for the purpose of understanding the method, it is best to use pencil and paper. Once you've mastered that, you can try to do the methods in your head.

Estimate: An estimate usually means a number that's easy to work with and close to the original number. There's no objective definition of an estimate. Whenever the word "estimate" or "nearby" is used, you can actually use any number. However, the method is usually easiest if the estimate is close to the original number.

Round Number: A round number usually means a number that has only one (or maybe two) non-zero digit. There's no objective definition of a round number. Whenever the phrase "round number" is used, you can actually use any number. However, the method is usually easiest if you use a number that only has one non-zero digit.

Difference: Whenever "difference" is mentioned, sign matters, unless it's the square of the difference (because a negative squared would be a positive) or the sign is included in a later step (meaning that a later step tells you whether to add or subtract). If it's the square of the difference or if the sign is included in a later step, the difference is always positive.

Equations: The algebraic equations are shown only to make sure you correctly understand the steps. It is easier to memorize the steps and it is not necessary to memorize the equations.

Place Values: Brackets are used to separate place values. The value of the bracket expression is the right-most place value, plus 10 times the next place value over to the left, plus 100 times the next, plus 1,000 times the next, etc. If each place value is a single-digit positive integer, the number means what it would mean without the brackets. For example, [6][4][7]=647. The brackets exist in case a place value is not a single-digit positive integer. If a place value is 10 or greater, what's before its last digit is added to the next higher place value. If a place value contains a decimal, what's after its decimal point is added to the next lower place value. If a place value contains a negative number, it is carried from the higher place value. For example, [5][17][2]=672, [8][-4][7]=767, and [3][4.5][3]=348.

Modulo: Modulo means divide and only keep the remainder. For example, if you divide 47 by 11, you'll get 4 with a remainder of 3, so 47 modulo 11=3. Sometimes, "modulo" is shortened to "mod".

Order Of Operations: This book follows the standard order of operations. Everything in parentheses is done first, then exponents, then multiplication and division, then addition and subtraction. If two operations are from the same category (multiplication and division are the same category, addition and subtraction are the same category), they are solved left to right.

Cross Multiply: Cross multiplying is multiplying one place value from one number by the other place value in the other number, and vice versa, then adding the two products. For example, "cross multiplying the tens and ones place" would mean the ones place from the first number times the tens place from the second number, plus the tens place from the first number times the ones place from the second number.

Equivalent Expression Rule: To make a math problem easier to do in your head, you can forget about the original expression as soon as you create an equivalent expression. For example, if you're using the Stretch Out Method to change 76·79 into 75·80+1·4, you can just pretend that 75·80+1·4 was the original problem.

CHAPTER

ADDITION, SUBTRACTION, AND COUNTING

Some of the addition and subtraction tricks in this chapter will be used a lot in later chapters when multiplying, since many multiplication tricks involve adding or subtracting several steps.

> **Subtracting Large Numbers** (and avoiding carrying mistakes)

When subtracting, you were most likely taught that if you're subtracting a higher digit from a lower digit, you need to carry. However, it's very easy to carry wrong if you're doing it in your head, so it's easiest if you change it into an expression that only involves subtracting from 0s.

To subtract from a set of zeros, subtract 1 from the place immediately before the 0s, subtract every digit except the last from 9, and subtract the last from 10. For example:

$430,000 - 7,253 = [43 - 1][9 - 7][9 - 2][9 - 5][10 - 3] = 422,747$

To make an expression that only subtracts from 0s, find the difference between the digits for each place value, and keep the difference in the number that had the higher digit. The other number will have a 0 in that place. For example

$$536,274 - 278,183 = 300,101 - 42,010$$

For the hundred thousands place, 5-2=3, and the first number had the higher digit in that place, so the 3 stays in the first number, while the second number has a 0 in that place. For the ten thousands place, 7-3=4 and the second number had the higher digit in that place, so the 4 stays in the second number, while the first number has a 0 in that place. Do the same thing for the rest of the digits.

Now you're only subtracting from 0s. Use the same method used before for subtracting from all 0s. Every time a zero comes immediately before a nonzero digit in the first number, separate the equation between the zero and the nonzero digit and calculate the smaller equations separately. In this example, you can do 300-42, 10-1, and 1-0 separately.

$$536,274 - 278,183 = 300,101 - 42,010$$
$$= [3 - 1][9 - 4][10 - 2][1 - 1][10 - 1][1 - 0] = 258,091$$

➢ **Adding Many Numbers** (by guessing the average)

It may take a while to add many numbers one at a time, but it can be made easier by guessing at the average, then adding just the differences from the average.

For example, to add the numbers:

82, 84, 85, 85, 87, 89, 91, 94

Pick a number somewhere in the middle of those, and subtract it from all of them. You could pick one of those numbers, so it cancels out. You could pick a number halfway between two of those numbers, so two of them cancel out. You could pick an easy number to multiply by. You could pick the smallest number to avoid negatives.

Subtracting 87 (to cancel out 85, 87, and 89), leaves

$-5, -3, -2, (-2), (0), (2), 4, 7$

The 2 and -2 cancel. Parentheses are shown around the numbers that cancel. Subtracting the negatives from the positives leaves

$(7 + 4) - (5 + 3 + 2) = 11 - 10 = 1$

The sum of the original numbers is how many numbers there were times the number subtracted from each, plus your result from adding the differences. For this example

$87 \cdot 8 + 1 = 697$

Since 85 is an easier number to multiply by 8, you may pick that instead. Subtracting 85 from each leaves

$-3, -1, (0), (0), 2, 4, 6, 9$

$(2 + 4 + 6 + 9) - (3 + 1) = 21 - 4 = 17$

$85 \cdot 8 + 17 = 697$

If you avoid negatives by choosing the lowest number to subtract

$(0), 2, 3, 3, 5, 7, 9, 12$

$2 + 3 + 3 + 5 + 7 + 9 + 12 = 41$

$82 \cdot 8 + 41 = 697$

This is useful when finding the average of multiple test or quiz grades. To figure out the average of several numbers, guess at the average, then subtract your guess from each number. Add the results, and divide that by how many numbers there are. Then, add your guessed average. Using the eight numbers given before

$82, 84, 85, 85, 87, 89, 91, 94$

Assume that you guessed an average of 85. Subtracting 85 from each leaves

$-3, -1, (0), (0), 2, 4, 6, 9$

10   How To Be A Math Genius

$$(2 + 4 + 6 + 9) - (3 + 1) = 21 - 4 = 17$$

So the average is

$$guessed\ average + \frac{total\ of\ differences}{how\ many\ numbers} = 85\frac{17}{8} = 87\frac{1}{8}$$

➤ **Adding Many Numbers** (rounding to the nearest 10)

Another option is to round each number to the nearest 10. First add all the nearest 10s, then add the differences. If two numbers end in 5, round one up and one down so the differences cancel. Using the same numbers from before,

$$80, 80, 80, 90, 90, 90, 90, 90$$

The sum of the nearest 10s is 690. Now, add the differences

$$2, 4, 5, -5, -3, -1, 1, 4$$

Those add to 7. Adding that to the sum of the nearest 10s results in 697, so the sum of the original numbers is 697.

➤ **Adding Many Single Digit Numbers** (adding 10 at a time)

If you're adding many single digit numbers, you can make it easier by separating them into combinations that equal something divisible by 10, remembering which number you're on and which numbers you've skipped. For example,

$$2 + 5 + 3 + 4 + 8 + 5 + 1 + 7 + 8 + 6 + 9 + 2 + 3 + 4 + 2 + 1 + 7 + 4$$
$$+ 8 + 6 + 5 + 2$$

2+5+3=10. After that, you can skip the 8 for now and add 4+5+1=10 to get 20. Add the skipped 8 to get 28. At this point, it's hard to use the trick because you'd have to skip four numbers to get the 2, so you can just add the 7 to get 35. You can now skip the 8 and add 6+9=15 to get 50, then add the skipped 8 to get 58. Add the next 2 to get 60. 3+4+2+1=10 to make 70. Skip the 7 and 8 and add 4+6=10 to make 80, then add the skipped 7 and 8 to make 95. Add the 5 to get 100 then add the 2 to get 102.

➤ **Adding Many Single Digit Numbers** (on a calculator and in your head)

If you're using a calculator to add many numbers, you can speed it up by adding a few at a time in your head. Using the previous example,

$$2 + 5 + 3 + 4 + 8 + 5 + 1 + 7 + 8 + 6 + 9 + 2 + 3 + 4 + 2 + 1 + 7 + 4$$
$$+ 8 + 6 + 5 + 2$$

If you add two numbers at a time in your head, you can add them quicker on a calculator.

$$7 + 7 + 13 + 8 + 14 + 11 + 7 + 3 + 11 + 14 + 7 = 102$$

➤ **Adding Many Single Digit Numbers** (putting equal numbers together)

If there are only a few different numbers, you can add each different number separately. For example,

$$7 + 4 + 6 + 7 + 7 + 4 + 7 + 6 + 4$$

There are only three different numbers (4, 6, 7). There are three 4s, two 6s, and four 7s.

$$7 + 4 + 6 + 7 + 7 + 4 + 7 + 6 + 4 = 3 \cdot 4 + 2 \cdot 6 + 4 \cdot 7 = 52$$

➤ **Counting**

Most people, when counting how many objects there are in a line, will count one by one. If you're good at judging where the middle is, you can count them quicker.

Number of Objects	Arrangement
3	1 in middle and 1 on each side
4	2 on each side
5	1 in middle and 2 on each side
6	3 on each side
7	1 in middle and 3 on each side
2n	n on each side
2n+1	1 in middle and n on each side

If the line is too large to judge the middle or to judge how many are on each side, you can count 5 at a time. Either make a mark every 5 or use your hand to keep track of your place. For example, it may be difficult to quickly count how many X's there are here.

X X X X X X X X X X X X X X X X X X X X X X X X X X X X X X X X X X X X X X X X X

But if you put a mark after every 5, it's easier to see that there are 41.

X X X X X|X X X X X|X X X X X|X X X X X|X X X X X|X X X X X|X X X X X|X X X X X|X

If you're counting a large quantity of an object and you're worried about losing your place if you count by 1, arrange them into 5x5 rectangles (25 in each) or 1-2-3-4 triangles (10 in each). For example, all three of the following photos have 67 dots, but it's easier to know that in the second and third photos than the first.

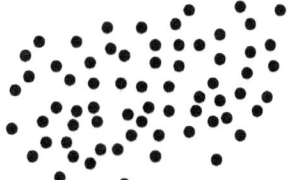 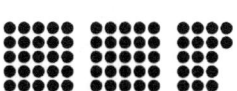 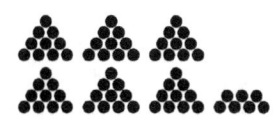

If you're counting a large quantity of an object and you can't see all of them at once, separate the objects into categories so that every object is in exactly one category. For example, imagine that you had to count how many boxes there are.

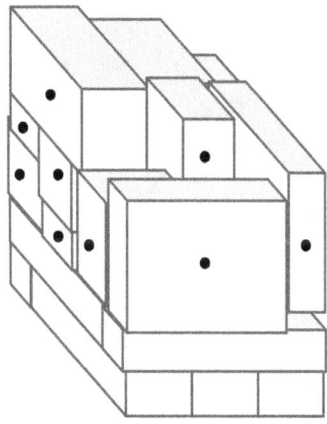

The first category should be the flat layers on the bottom, which are easy to count because they're neatly arranged. Assume that you know that there are 2 flat layers of 4 boxes on the bottom. That's 8 boxes.

The next category should be the boxes that are visible from two adjacent sides and not in the flat layers. We won't count the boxes visible only from the top, because you wouldn't actually be able to see them if you were looking at actual boxes. The boxes marked with a dot are visible from the two sides shown. There are 9.

The last category should be the boxes that are in neither of the other two categories, meaning that they are not in the flat layers and are not visible from the two sides shown. You can't see them in the picture, but let's assume that there are 4. The total number of boxes would be 8+9+4=21.

CHAPTER

12 MULTIPLICATION TABLE

In elementary school, you're most likely required to memorize the products of any two numbers between 1 and 12. This chapter will show you what to do if you forget one of the products. The methods that are shown here will be used for larger numbers in the other chapters, with each method having its own chapter.

	1	2	3	4	5	6	7	8	9	10	11	12
1	1	2	3	4	5	6	7	8	9	10	11	12
2	2	4	6	8	10	12	14	16	18	20	22	24
3	3	6	9	12	15	18	21	24	27	30	33	36
4	4	8	12	16	20	24	28	32	36	40	44	48
5	5	10	15	20	25	30	35	40	45	50	55	60
6	6	12	18	24	30	36	42	48	54	60	66	72
7	7	14	21	28	35	42	49	56	63	70	77	84
8	8	16	24	32	40	48	56	64	72	80	88	96
9	9	18	27	36	45	54	63	72	81	90	99	108
10	10	20	30	40	50	60	70	80	90	100	110	120
11	11	22	33	44	55	66	77	88	99	110	121	132
12	12	24	36	48	60	72	84	96	108	120	132	144

If you're multiplying by 10, just put a 0 on the end. If you're multiplying an even number by 5, divide by 2 and put a 0 on the end. If you're multiplying an odd number by 5, subtract 1, divide by 2, and put a 5 on the end.

For each of the following methods, it's easier to memorize the listed steps than to memorize the algebraic equations. The equations are there only so you can see if you understand the steps correctly.

➢ **Square Estimation Method**

The most important products to memorize in the table are the squares, because knowing the squares will allow you to use the Difference of Squares Method. The best way to square a number is to use the Square Estimation Method.

Step 1: Pick a nearby number to estimate.
Step 2: Subtract half the estimate from the original number.
Step 3: Multiply the result from Step 2 by twice the estimate.
Step 4: Add the square of the difference between the original number and the estimate to the result from Step 3.

For now, the estimate should be 5 or 10. Bigger numbers, which will be used in other chapters, will use bigger estimates. If your estimate is 5, you can subtract 3 (instead of subtracting 2.5) for Step 2, and multiply by 10 and add 5 (instead of just multiplying by 10) for Step 3. The estimate can actually be anything, but it's easiest if you pick a number that's close to your original number and easy to multiply by. Here are some examples of the Square Estimation Method.

Original Number: 4
Step 1: 5
Step 2: $4 - 3 = 1$
Step 3: $1 \cdot 10 + 5 = 15$
Step 4: $15 + 1 = 16$

Original Number: 7
Step 1: 5 *or* 10
Step 2: $7 - 3 = 4$ *or* $7 - 5 = 2$
Step 3: $4 \cdot 10 + 5 = 45$ *or* $2 \cdot 20 = 40$
Step 4: $45 + 4 = 49$ *or* $40 + 9 = 49$

Original Number: 12

16 How To Be A Math Genius

Step 1: 10
Step 2: 12 – 5 = 7
Step 3: 7·20 = 140
Step 4: 140 + 4 = 144

> ➤ **Square Estimation Method:** (equation form)

If "a" is your original number and "x" is your estimate:

$$a^2 = \left(a - \frac{x}{2}\right)2x + (a - x)^2$$

Or if your estimate is odd and you want to only use whole numbers:

$$a^2 = \left(a - \frac{x + 1}{2}\right)2x + x + (a - x)^2$$

> ➤ **Square Estimation Method** (alternative steps)

You can slightly change the steps to do it a different way.

Step 1: Pick a nearby number to estimate.
Step 2: Double your original number.
Step 3: Subtract the estimate from your result from Step 2, then multiply by the estimate.
Step 4: Add the square of the difference between the original number and the estimate to the result from Step 3.

Original Number: 4
Step 1: 5
Step 2: 4·2 = 8
Step 3: 5(8 – 5) = 15
Step 4: 15 + 1 = 16

Original Number: 7
Step 1: 5 *or* 10
Step 2: 7·2 = 14
Step 3: 5(14 – 5) = 45 *or* 10(14 – 10) = 40
Step 4: 45 + 4 = 49 *or* 40 + 9 = 49

Original Number: 12
Step 1: 10

Step 2: $12 \cdot 2 = 24$
Step 3: $10(24 - 10) = 140$
Step 4: $140 + 4 = 144$

➤ **Square Estimation Method** (alternative steps, equation form)

If "a" is your original number and "x" is your estimate:

$$a^2 = (2a - x)x + (a - x)^2$$

➤ **Stretch Out Method**

The Square Estimation Method only works when squaring. If you're multiplying two different numbers, you can use the Stretch Out Method, which is usually the best method for multiplying two close numbers.

Step 1: Make the numbers easier to multiply by adding something to one number and subtracting the same thing from the other number. The resulting numbers are your estimates.
Step 2: Multiply the two estimates.
Step 3: Find the difference between either one original number and each estimate, or one estimate and each original number.
Step 4: Multiply the two differences.
Step 5: If the original numbers are closer to each other than the estimates are to each other, add Steps 2 and 4. If the original numbers are further away, subtract Steps 2 and 4.

For Step 3, be careful to avoid finding the difference between the two original numbers or between the two estimates, as those are not used. One of the two differences in Step 3 will always be the number you added and subtracted in Step 1. For now, one of the estimates will always be 5 or 10. Bigger numbers will be used in other chapters. Here are some examples.

Original Numbers: $7 \cdot 8$
Step 1: $7 - 2 = 5, 8 + 2 = 10$
Step 2: $5 \cdot 10 = 50$
Step 3: $10 - 8 = 2 \ or \ 7 - 5 = 2; \ 10 - 7 = 3 \ or \ 8 - 5 = 3$
Step 4: $2 \cdot 3 = 6$
Step 5: *Original numbers closer*. $50 + 6 = 56$

Original Numbers: $6 \cdot 12$

Step 1: $6 + 2 = 8$, $12 - 2 = 10$
Step 2: $8 \cdot 10 = 80$
Step 3: $12 - 10 = 2$ *or* $8 - 6 = 2$; $12 - 8 = 4$ *or* $10 - 6 = 4$
Step 4: $2 \cdot 4 = 8$
Step 5: *Original numbers further.* $80 - 8 = 72$

Original Numbers: $7 \cdot 16$
Step 1: $7 + 3 = 10$, $16 - 3 = 13$
Step 2: $10 \cdot 13 = 130$
Step 3: $16 - 13 = 3$ *or* $10 - 7 = 3$; $16 - 10 = 6$ *or* $13 - 7 = 6$
Step 4: $3 \cdot 6 = 18$
Step 5: *Original numbers further.* $130 - 18 = 112$

➢ **Stretch Out Method** (equation form)

If "a" and "b" are your original numbers and "x" is the amount added or subtracted to create your estimates:

$$ab = (a + x)(b - x) + x(a + x - b)$$

"a+x-b" is the difference between one original number and the other estimate.

➢ **Stretch Out Method: Single Estimate Variation**

The Single Estimate Method is just like the Stretch Out Method, except that you only focus on one estimate (usually a round number) and the other estimate is implied. Like the Stretch Out Method, it's best for multiplying two close numbers.

Step 1: Pick a nearby number as an estimate.
Step 2: Subtract the estimate from each number (sign matters). Your new numbers are called the "differences".
Step 3: Square the estimate.
Step 4: Add the two differences and multiply by the estimate.
Step 5: Multiply the differences.
Step 6: Add the results of Steps 3 through 5.

Your estimate for Step 1 can actually be any number, but it's easiest if you pick a nearby round number, most likely 10 for this chapter. Here are some examples:

Original Numbers: $7 \cdot 8$

Step 1: 10
Step 2: 7 − 10 =− 3; 8 − 10 =− 2
Step 3: 100
Step 4: 10(− 3 +− 2) =− 50
Step 5: − 2· − 3 = 6
Step 6: 100 − 50 + 6 = 56

Original Numbers: 6·12
Step 1: 10
Step 2: 6 − 10 =− 4; 12 − 10 = 2
Step 3: 100
Step 4: 10(− 4 + 2) =− 20
Step 5: − 4·2 =− 8
Step 6: 100 − 20 − 8 = 72

Original Numbers: 7·16
Step 1: 10
Step 2: 7 − 10 =− 3; 16 − 10 = 6
Step 3: 100
Step 4: 10(− 3 + 6) = 30
Step 5: − 3·6 =− 18
Step 6: 100 + 30 − 18 = 112

➢ **Stretch Out Method: Single Estimate Variation** (equation form)

If "a" and "b" are your original numbers and "x" is your estimate:

$$ab = x^2 + x((a - x) + (b - x)) + (a - x)(b - x)$$

➢ **Stretch Out Method: Single Estimate Variation** (alternate steps)

Here's a different way to do the Single Estimate Method.

Step 1: Pick a nearby round number as an estimate.
Step 2: Add the two original numbers.
Step 3: Subtract your estimate from the result from Step 2, then multiply by your estimate.
Step 4: Add, to the result from Step 3, the product of the differences (sign matters) from your estimate.

Using the same examples as before:

Original Numbers: 7·8
Step 1: 10
Step 2: $7 + 8 = 15$
Step 3: $10(15 - 10) = 50$
Step 4: $50 + (7 - 10)(8 - 10) = 56$

Original Numbers: 6·12
Step 1: 10
Step 2: $6 + 12 = 18$
Step 3: $10(18 - 10) = 80$
Step 4: $80 + (6 - 10)(12 - 10) = 72$

Original Numbers: 7·16
Step 1: 10
Step 2: $7 + 16 = 23$
Step 3: $10(23 - 10) = 130$
Step 4: $130 + (7 - 10)(16 - 10) = 112$

> ➤ **Stretch Out Method: Single Estimate Variation** (alternate steps, equation form)

If "a" and "b" are your original numbers and "x" is your estimate:

$$ab = (a + b - x)x + (a - x)(b - x)$$

> ➤ **Stretch Out Method: Difference Of Squares Variation**

If you're multiplying two numbers that are either both odd or both even, and you already have the squares memorized, you can use the Difference Of Squares Method. You can actually do it with one even and one odd, but the midpoint will not be a whole number, so it will be harder to do.

Step 1: Find the midpoint and square it.
Step 2: Find the difference between the midpoint and the original numbers and square it.
Step 3: Subtract the two squares.

This is exactly like the Stretch Out Method with both estimates being the same number. Here are some examples.

Original Numbers: 7·9

Step 1: $8^2 = 64$
Step 2: $1^2 = 1$
Step 3: $64 - 1 = 63$

Original Numbers: $8 \cdot 14$
Step 1: $11^2 = 121$
Step 2: $3^2 = 9$
Step 3: $121 - 9 = 112$

➢ **Stretch Out Method: Difference Of Squares Variation** (equation form)

If "a" and "b" are your original numbers:

$$ab = \left(\frac{a+b}{2}\right)^2 - \left(\frac{a-b}{2}\right)^2$$

(a+b)/2 is the midpoint, and (a-b)/2 is the distance from the midpoint.

CHAPTER

2'S AND 5'S

Because 10 is the common base used when expressing numbers, it's usually relatively easy to multiply by 2 or by 5.

To multiply by 5, multiply by 10 then divide by 2. To divide by 5, multiply by 2 then divide by 10.

Doubling and halving can be useful for methods explained in other chapters. For example, if you're using the Stretch Out Method (which was briefly mentioned in Chapter 2 and will be explained in more detail in Chapter 6), it's best if you're multiplying numbers that are close together. If one of the numbers is close to twice the other, you can double the lesser number or halve the greater number to make them closer, then do the opposite at the end.

➢ **Multiplying by 2**

To multiply a number by 2, you can do each digit one at a time. Multiply each digit by 2, subtract 10 if your result is 10 or greater, and add 1 if the next digit is 5 or greater. If the first digit is 5 or greater, the solution will start with a carried 1.

For example, assume you were given the number 394,872.

- The hundred-thousands digit in your solution is 7, because 3·2=6 and you add 1 because the next digit (9) is 5 or greater.
- The ten-thousands digit in your solution is 8, because 9·2=18 (subtract 10) and you do not add 1 because the next digit (4) is less than 5.
- The thousands digit in your solution is 9 because 4·2=8 and you add 1 because the next digit (8) is 5 or greater.
- The hundreds digit in your solution is 7, because 8·2=16 (subtract 10) and you add 1 because the next digit (7) is 5 or greater.
- The tens digit in your solution is 4, because 7·2=14 (subtract 10) and you do not add 1 because the next digit (2) is less than 5.
- The ones digit in your solution is 4 because 2·2=4 and you do not add 1 because there's no next digit.

394,872·2=789,744.

➢ **Dividing by 2**

To divide a number by 2, you can do each digit one at a time. Divide each digit by 2, round down to the nearest whole number, and add 5 if the previous digit is odd. If the last digit is odd, the solution will have a 5 after the decimal point.

For example, assume you were given the number 394,872.

- The hundred-thousands digit in your solution is 1, because 3/2=1.5 (round down) and you do not add 5 because there is no previous digit.
- The ten-thousands digit in your solution is 9 because 9/2=4.5 (round down) and you add 5 because the previous digit (3) is odd.
- The thousands digit in your solution is 7, because 4/2=2 and you add 5 because the previous digit (9) is odd.
- The hundreds digit in your solution is 4 because 8/2=4 and you do not add 5 because the previous digit (4) is even.
- The tens digit in your solution is 3 because 7/2=3.5 (round down) and you do not add 5 because the previous digit (8) is even.
- The ones digit in your solution is 6 because 2/2=1 and you add 5 because the previous digit (7) is odd.

394,872/2=197,436

> ➤ **Multiplying By A Number Divisible By 5**

When you're multiplying two or more numbers, you can factor by multiplying one of the numbers by something and dividing another number by the same thing. The product will not change. You can also multiply or divide one of the numbers, find the product of your new numbers, then undo the factoring at the end.

When multiplying by a number ending in 5, double the number ending in 5 and halve the other number.

$85 \cdot 12 = 170 \cdot 6 = 1,020$
$65 \cdot 34 = 130 \cdot 17 = 2,210$
$15 \cdot 78 = 30 \cdot 39 = 1,170$

If it ends in 25 or 75, quadruple it and divide the other number by 4.

$25 \cdot 32 = 100 \cdot 8 = 800$
$75 \cdot 92 = 300 \cdot 23 = 6,900$

If the other number is not divisible by 2 (or 4 for numbers ending in 25 or 75), take out the remainder and multiply it separately.

$75 \cdot 49 = 75(48 + 1) = 300 \cdot 12 + 75 = 3,600 + 75 = 3,675$
$35 \cdot 83 = 35(82 + 1) = 70 \cdot 41 + 35 = 2,870 + 35 = 2,905$

You can continue that pattern. If it ends in something divisible by 125, you can multiply it by 8 to make it divisible by 1,000, and divide the other number by 8.

$125 \cdot 72 = 1,000 \cdot 9 = 9,000$
$375 \cdot 56 = 3,000 \cdot 7 = 21,000$

When multiplying by a multiple of 10, remove all the 0s at the end of the numbers, then put them on when you're done.

$60 \cdot 700 = 6 \cdot 7 \cdot 1,000 = 42,000$
$1,200 \cdot 9,000 = 12 \cdot 9 \cdot 100,000 = 10,800,000$

CHAPTER

STANDARD METHOD

The Standard Method involves multiplying each digit in one number by each digit in the other number. It is most likely the easiest method to understand and the easiest method to do on pencil and paper. The only drawback is that it may involve a lot of carrying, which may be difficult to do in your head.

Step 1: Reverse the order of the digits in one of the numbers (so it reads right to left, the value of the number does not actually change), and place the numbers so the highest valued digits are lined up.

Step 2: Multiply any digits that are lined up. This will be the highest place value of your answer.

Step 3: Slide one of the numbers over one place so the highest valued digit of one number is lined up with the second highest digit of the other number, and repeat Step 2. More than one set of digits is together, so add the products of each set. This will be the next lower place value. If the result is 10 or greater, carry into the previous digit.

Step 4: Repeat Step 3 until only the lowest valued digits are lined up. Carry each time if a place value is 10 or greater.

Whenever doing the Standard Method or any variations of it, always imagine that there are an infinite number of invisible zeros on each side, and that the method is complete when only the invisible zeros are lined up.

Here are some examples.

Original Numbers: 124·347

Step 1: $\begin{bmatrix} 421 \\ 347 \end{bmatrix}$

Step 2: $\begin{bmatrix} 421 \\ 347 \end{bmatrix} = 1 \cdot 3 = 3$

Step 3: $\begin{bmatrix} 421 \\ 347 \end{bmatrix} = 2 \cdot 3 + 1 \cdot 4 = 10$

Step 4: $\begin{bmatrix} 421 \\ 347 \end{bmatrix}\begin{bmatrix} 421 \\ 347 \end{bmatrix}\begin{bmatrix} 421 \\ 347 \end{bmatrix} = [4 \cdot 3 + 2 \cdot 4 + 1 \cdot 7][4 \cdot 4 + 2 \cdot 7][4 \cdot 7] = [27][30][28]$

Step 5: $[3][10][27][30][28] = 43,028$

It's easier if you carry as soon as you get a new place value, instead of waiting until the end. In the previous example, it's easier to just remember 3, then 40, then 427, then 4300, then 43028.

The Standard Method works best when one of the numbers has all low digits.

Original Numbers: 112·673

Step 1: $\begin{bmatrix} 211 \\ 673 \end{bmatrix}$

Step 2: $\begin{bmatrix} 211 \\ 673 \end{bmatrix} = 1 \cdot 6 = 6$

Step 3: $\begin{bmatrix} 211 \\ 673 \end{bmatrix} = 1 \cdot 6 + 1 \cdot 7 = 13$

Step 4: $\begin{bmatrix} 211 \\ 673 \end{bmatrix}\begin{bmatrix} 211 \\ 673 \end{bmatrix}\begin{bmatrix} 211 \\ 673 \end{bmatrix} = [2 \cdot 6 + 1 \cdot 7 + 1 \cdot 3][2 \cdot 7 + 1 \cdot 3][2 \cdot 3] = [22][17][6]$

Step 5: $[6][13][22][17][6] = 75,376$

It's easier if you carry as soon as you get a new place value, instead of waiting until the end. In the previous example, it's easier to just remember 6, then 73, then 752, then 7537, then 75376.

➢ **Standard Method** (equation form)

$[a][b][c] \cdot [d][e][f] = [ad][ae + bd][af + be + cd][bf + ce][cf]$

➢ **Standard Method: Redistributed Place Values**

If some of the digits are high, you can redistribute by subtracting 10 from the place value with a high digit then adding 1 to the next higher place value. For this example, 189 is redistributed by subtracting 10 from the ones place and adding 1 to the tens place to get [1][9][-1], then by subtracting 10 from the tens place and adding 1 to the hundreds place to get [2][-1][-1]. You can even redistribute both numbers. If you redistribute, you must keep the same distribution for the entire method. You can't sometimes use 189 and sometimes use [2][-1][-1]. This can make it easier to do in your head, but there's usually no reason to do this if you're using pencil and paper.

Original Numbers: 189·673

Step 1: $\begin{bmatrix} [-1][-1][2] \\ [6][7][3] \end{bmatrix}$

Step 2: $\begin{bmatrix} [-1][-1][2] \\ [6][7][3] \end{bmatrix} = 2 \cdot 6 = 12$

Step 3: $\begin{bmatrix} [-1][-1][2] \\ [6][7][3] \end{bmatrix} = -1 \cdot 6 + 2 \cdot 7 = 8$

Step 4: $\begin{bmatrix} [-1][-1][2] \\ [6][7][3] \end{bmatrix}\begin{bmatrix} [-1][-1][2] \\ [6][7][3] \end{bmatrix}\begin{bmatrix} [-1][-1][2] \\ [6][7][3] \end{bmatrix}$

$= [-1 \cdot 6 + -1 \cdot 7 + 2 \cdot 3][-1 \cdot 7 + -1 \cdot 3][-1 \cdot 3] = [-7][-10][-3]$

Step 5: $[12][8][-7][-10][-3] = 127,197$

If you carry after each digit, you'll get 12, then 128, then 1273, then 12720, then 127197.

If one of the digits is close to 5, you could use a fractional place value. For example, 941 could be redistributed by subtracting 5 from the tens place and adding 0.5 to the hundreds place to get [9.5][-1][1], then by subtracting 10 from the hundreds place and adding 1 to the thousands place to get [1][-0.5][-1][1]. Using fractional place values may be hard to understand (so you can skip this method), but if you understand it, it may make some problems easier to do in your head due to less carrying.

$941 \cdot 638 = [1]\left[-\frac{1}{2}\right][-1][1] \cdot 638$

$= [6 \cdot 1]\left[3 \cdot 1 + 6 \cdot -\frac{1}{2}\right]\left[8 \cdot 1 + 3 \cdot -\frac{1}{2} + 6 \cdot -1\right]\left[8 \cdot -\frac{1}{2} + 3 \cdot -1 + 6 \cdot 1\right]$

$[8 \cdot -1 + 3 \cdot 1][8 \cdot 1]$

$= [6][3 - 3][8 - 1.5 - 6][-4 - 3 + 6][-8 + 3][8]$

$= [6][0][0.5][-1][-5][8] = 600,358$

➢ **Standard Method: All At Once**

You can also multiply a number all at once by each digit in the other number, which can be possible to do in your head if one of the numbers is small.

$13 \cdot 263 = [13 \cdot 2][13 \cdot 6][13 \cdot 3] = [26][78][39] = 3,419$

Carrying after each place value would give you 26, 338, 3419.

➢ **Standard Method: Special Cases**

Because of the Standard Method, some numbers can be easy to multiply using the sums or differences of place values.

● Multiplying By 9

The number 9 can be redistributed as [1][-1]. To multiply by 9, take the difference between each digit and the digit before it (sign matters). You'll always subtract for the last digit, because it will always be an invisible zero minus your ones place.

$38,492 \cdot 9 = [3][8 - 3][4 - 8][9 - 4][2 - 9][- 2]$
$= [3][5][- 4][5][- 7][- 2] = 346,428$

You can use the subtraction method shown in Chapter 1.

$[3][5][- 4][5][- 7][- 2] = 350,500 - 4,072 = 346,428$

If you're multiplying by a number divisible by 9, factor out the 9 then multiply by 9 at the end. In this example, you can factor out the 9 from 54 to get 6, then multiply by 9 after you've already multiplied $231 \cdot 6$.

$231 \cdot 54 = 231 \cdot 6 \cdot 9 = 1,386 \cdot 9 = [1][2][5][- 2][- 6] = 12,474$

● Multiplying By 11

To multiply by 11, take the sum of every two consecutive digits.

$38,492 \cdot 11 = [3][3 + 8][8 + 4][4 + 9][9 + 2][2] = [3][11][12][13][11][2] = 423,412$

If you're multiplying by a number divisible by 11, factor out the 11 then multiply by 11 at the end. In this example, you can factor out the 11 from 77 to get 7, then multiply by 11 after you've already multiplied 173·7.

$$173 \cdot 77 = 173 \cdot 7 \cdot 11 = 1{,}211 \cdot 11 = 13{,}321$$

- Multiplying By 111

Multiplying by 11 involved the sum of every two digits. Multiplying by 111 involves the sum of every three. Remember the invisible zeros on each side.

$$38{,}492 \cdot 111$$

$$[3][3+8][3+8+4][8+4+9][4+9+2][9+2][2]$$
$$= [3][11][15][21][15][11][2] = 4{,}272{,}612$$

This can be useful when multiplying by 37, because you can first divide by 3 then multiply by 111, because 37·3=111.

- Multiplying By Other Special Case Numbers

The Standard Method is also useful for multiplying by 89 (each digit minus the two before it), 889 (each digit minus the three before it), 911 (each digit minus the one before it and plus the next two), and any other number entirely made of place values of 1, 0, and -1.

$$346 \cdot 89 = 346 \cdot [1][-1][-1] = [3][4-3][6-4-3][-6-4][-6]$$
$$= [3][1][-1][-10][-6] = 30{,}794$$

$$346 \cdot 889 = 346 \cdot [1][-1][-1][-1]$$
$$= [3][4-3][6-4-3][-6-4-3][-6-4][-6]$$
$$= [3][1][-1][-13][-10][-6] = 307{,}594$$

$$346 \cdot 911 = 346 \cdot [1][-1][1][1]$$
$$= [3][4-3][6-4+3][-6+4+3][6+4][6]$$
$$= [3][1][5][1][10][6] = 315{,}206$$

➢ **Standard Method: Straight Down Variation**

For multiplying two 2-digit numbers, an easier way to do the Standard Method is to do the Straight Down Method.

Step 1: Multiply only the tens place by the tens place and the ones place by the ones place, also known as multiplying "straight down".

Step 2: Cross multiply, multiply by 10, and add to your result from Step 1.

Here are some examples.

Original Numbers: 64·87
Step 1: 6·8 = 48; 4·7 = 28; 64·87 = 4,828 + ...
Step 2: 6·7 + 4·8 = 74; 4,828 + 740 = 5,568

Original Numbers: 32·74
Step 1: 3·7 = 21; 2·4 = 8; 32·74 = 2,108 + ...
Step 2: 3·4 + 2·7 = 26; 2,108 + 260 = 2,368

> ➤ **Standard Method: Last Two Digits Only**

If you're going to be using a different method because the Standard Method is too difficult for the problem you're solving, it may help to at least figure out the last two digits.

The last digit is the ones digit of the product of the ones places. The second last digit is the tens digit of the product of the ones places, plus the ones digits of the cross-multiplied products of the ones and tens places.

For example, 378·437

Last Digit: 8·7 = 5[6]
Second Last Digit: 8·7 = [5]6; 7·7 = 4[9]; 8·3 = 2[4]; 5 + 9 + 4 = 1[8]
Last Two Digits: 86

> ➤ **Standard Method: Multiplying By Slightly Less Than A Round Number**

You've most likely learned this as the "Distributive Property". If you're multiplying by something slightly less than a round number, distribute it as the round number minus the difference.

7·892 = 7(900 − 8) = 6,300 − 56 = 6,244
6·497 = 6(500 − 3) = 3,000 − 18 = 2,982

➤ Standard Method: Straight Down Variation For Three Digits

The Straight Down Method is slightly different for three digit numbers. There are two ways to do it. You could either keep the three place values, or combine the tens and ones to make two place values.

If you're keeping three place values:

> **Step 1:** Multiply each place value straight down, giving 2 place values to each in the solution.
> **Step 2:** Cross multiply the hundreds and tens places, and add to the thousands place of your solution.
> **Step 3:** Cross multiply the hundreds and ones places, and add to the hundreds place of your solution.
> **Step 4:** Cross multiply the tens and ones places, and add to the tens place of your solution.

Original Numbers: 332·749
Step 1: 3·7 = 21; 3·4 = 12; 2·9 = 18; [21][0][12][0][18] = 211,218
Step 2: 3·4 + 3·7 = 33; 211,218 + 33,000 = 244,218
Step 3: 3·9 + 2·7 = 41; 244,218 + 4,100 = 248,318
Step 4: 3·9 + 2·4 = 35; 248,318 + 350 = 248,668

If you're combining the tens and ones place:

> **Step 1:** Multiply the hundreds place straight down, and the combined tens and ones place straight down. The tens and ones place product will take up 4 place values.
> **Step 2:** Cross multiply the hundreds place and the combined tens and ones place, and add to the hundreds place of your solution.

Original Numbers: 332·749
Step 1: 3·7 = 21; 32·49 = 1,568; [21][0][0][0][1,568] = 211,568
Step 2: 3·49 + 32·7 = 371; 211,568 + 37,100 = 248,668

CHAPTER

5

SQUARES

Many methods of multiplying two numbers involve squares. This chapter will involve squaring numbers up to 3 digits, and any methods involving squares.

> **Square Estimation Method**

Step 1: Pick a nearby number to estimate.
Step 2: Subtract half the estimate from the original number.
Step 3: Multiply the result from Step 2 by twice the estimate.
Step 4: Add the square of the difference between the original number and the estimate to the result from Step 3.

As shown in Chapter 2, for Steps 2 and 3 you can instead double the original number then subtract and multiply by the estimate.

When squaring a number greater than 25, the estimate for Step 1 usually should be the nearest 50. This will be easier if you have the squares of all numbers 1 through 25 memorized.

Original Number: 38
Step 1: 50

Step 2: $38 - 25 = 13$
Step 3: $13 \cdot 100 = 1{,}300$
Step 4: $12^2 = 144$; $1{,}300 + 144 = 1{,}444$

Original Number: 79
Step 1: 100
Step 2: $79 - 50 = 29$
Step 3: $29 \cdot 200 = 5{,}800$
Step 4: $21^2 = 441$; $5{,}800 + 441 = 6{,}241$

Original Number: 647
Step 1: 650
Step 2: $647 - 325 = 322$
Step 3: $322 \cdot 1{,}300 = 418{,}600$
Step 4: $3^2 = 9$; $418{,}600 + 9 = 418{,}609$

Original Number: 869
Step 1: 850
Step 2: $869 - 425 = 444$
Step 3: $444 \cdot 1{,}700 = 754{,}800$
Step 4: $19^2 = 361$; $754{,}800 + 361 = 755{,}161$

➤ **Square Estimation Method** (square estimate first)

If multiplying by twice the estimate is too hard, there are other options.

Multiplying half the estimate by twice the estimate results in the estimate squared. You can then find the difference from there. Start with the estimate squared, then add twice the estimate times the difference from the estimate. That will be your result for Step 3. In the previous two examples:

Step 1: Pick a nearby number to estimate.
~~Step 2: Subtract half the estimate from the original number.~~
~~Step 3: Multiply the result from Step 2 by twice the estimate.~~
Replacement Step: Square the estimate, then add twice the estimate times the difference from the estimate. Sign matters.
Step 4: Add the square of the difference between the original number and the estimate to the result from Step 3.

Original Number: 647
Step 1: 650

Replacement Step: $650^2 + 2 \cdot 650(647 - 650) = 422,500 - 3,900$
$= 418,600$
Step 4: $3^2 = 9$; $418,600 + 9 = 418,609$

Original Number: 869
Step 1: 850
Replacement Step: $850^2 + 2 \cdot 850(869 - 850) = 722,500 + 32,300$
$= 754,800$
Step 4: $19^2 = 361$; $754,800 + 361 = 755,161$

> ➤ **Square Estimation Method** (estimate so Step 2 is close to a multiple of 100)

You can use a different estimate, so the result from Step 2 (original number minus half the estimate) will be close to a multiple of 100. To do this, round the original number to the nearest 25. Your estimate will be the nearest multiple of 50 that, when divided by 2, ends in the same two digits as the nearest 25. If you're squaring a large number in Step 4, you can repeat the entire method using that number. In the previous two examples:

Original Number: 647
Nearest 25 = 650.
Nearest 50 *that ends in* 50 *when halved* = 700
Step 1: 700
Step 2: 647 − 350 = 297
Step 3: 297·1,400 = 415,800
Step 4: $53^2 = 100(53 - 25) + 9 = 2,809$; $415,800 + 2,809 = 418,609$

Original Number: 869
Nearest 25 = 875.
Nearest 50 *that ends in* 75 *when halved* = 950
Step 1: 950
Step 2: 869 − 475 = 394
Step 3: 394·1,900 = 748,600
Step 4: $81^2 = 200(81 - 50) + 361 = 6,561$; $748,600 + 6,561 = 755,161$

> ➤ **Square Estimation Method** (estimate 500 or 1,000)

Sometimes, it may be easier to estimate to 500 or 1,000, even if the original number is far away, because Step 3 would be easy (multiplying by 1,000 or

2,000) and would minimize carrying in Step 4. If you're squaring a large number in Step 4, you can repeat the entire method using that number.

Original Number: 647
Step 1: 500
Step 2: 647 – 250 = 397
Step 3: 397·1,000 = 397,000
Step 4: $147^2 = 300(147 - 75) + 9 = 21,609$;
397,000 + 21,609 = 418,609

Original Number: 869
Step 1: 1,000
Step 2: 869 – 500 = 369
Step 3: 369·2,000 = 738,000
Step 4: $131^2 = 300(131 - 75) + 361 = 17,161$;
738,000 + 17,161 = 755,161

➢ **Squares Ending In 5**

If you're squaring a number ending in 5, you can use a much quicker method than the Square Estimation Method. To square a number ending in 5, use the Stretch Out Method (see Chapter 6) by estimating to the next higher and next lower 10. The distance from each estimate will always be 5 and the original numbers will always be closer, so you'll always add 25.

$35^2 = 30·40 + 5·5 = 1,200 + 25 = 1,225$

$75^2 = 70·80 + 5·5 = 5,600 + 25 = 5,625$

$115^2 = 110·120 + 5·5 = 13,200 + 25 = 13,225$

If a number ends in 25 or 75, you can use the Stretch Out Method by estimating to the next higher and next lower 50. The distance from each estimate will always be 25 and the original numbers will always be closer, so you'll always add 625 (25·25).

$275^2 = 250·300 + 25·25 = 75,000 + 625 = 75,625$

$425^2 = 400·450 + 25·25 = 180,000 + 625 = 180,625$

Sometimes, if the nearest 50 is hard to do in your head, you can instead use the next 50. You'll add 5,625 (75·75) because the distance will be 75 instead of 25.

$875^2 = 850 \cdot 900 + 25 \cdot 25 = 765,000 + 625 = 765,625$

$875^2 = 800 \cdot 950 + 75 \cdot 75 = 760,000 + 5,625 = 765,625$

In that example, the second one is easier to do in your head because 800·950 can be factored by halving 800 and doubling 950 to get 400·1,900, so both are divisible by 100.

If a number ends in 15, 35, 65, or 85, you can stretch out by 15 to make one of the numbers divisible by 50. You'll add 225 (15·15) at the end.

$435^2 = 420 \cdot 450 + 15 \cdot 15 = 189,000 + 225 = 189,225$

Now that you know how to square a number, you can use methods that involve squares.

➢ **Difference Of Squares Method**

As done in Chapter 2, if you're multiplying two numbers that are both odd or both even, you can use the Difference Of Squares Method. Here are the steps.

Step 1: Find the midpoint and square it.
Step 2: Find the difference between the midpoint and the original numbers and square it.
Step 3: Subtract the two squares.

If it's hard to figure out the midpoint, the midpoint is always half the sum of the two original numbers, and the distance from the midpoint is always half the difference between the two original numbers. Here are some examples.

Original Numbers: 17·21
Step 1: $19^2 = 361$
Step 2: $2^2 = 4$
Step 3: $361 - 4 = 357$

Original Numbers: 36·52
Step 1: $44^2 = 1,936$
Step 2: $8^2 = 64$

Step 3: 1,936 − 64 = 1,872

If they're not close together but their difference is divisible by 20, this method may still be easy to do because the distance from the midpoint will be divisible by 10.

Original Numbers: 57·117
Step 1: $87^2 = 7,569$
Step 2: $30^2 = 900$
Step 3: 7,569 − 900 = 6,669

Original Numbers: 29·69
Step 1: $49^2 = 2,401$
Step 2: $20^2 = 400$
Step 3: 2,401 − 400 = 2,001

➢ **Sum Of Squares Method**

This method is not used too often, because it's not usually easy to quickly know if two numbers are both the sum of two squares. But if you're good with squares and you know that both numbers you're multiplying are the sum of two squares, this method can be useful.

If you're multiplying two numbers that are both the sum of two different squares, the product will be the sum of two squares in two different ways.

If a number is the sum of two squares in more than one way (65=49+16=64+1), either way works in Step 1.

Step 1: For each of the two numbers you're multiplying, find two numbers whose squares added together equal the number you're multiplying. The "first set" is the two numbers whose squares add to the first number, and the "second set" is the two numbers whose squares add to the second number.

Step 2: Multiply any number from one set by any number from the other set. Do the same with the other number in each set.

Step 3: Multiply any number from one set by the number in the other set that you didn't multiply it by in Step 2. Do the same with the other number in each set.

Step 4: The product will be the sum of two squares. One of those numbers being squared is the sum or difference of your results from Step 2. The other number being squared is the sum or difference of your results from Step 3. Your choices are either to add in Step 2 and subtract in Step 3, or vice versa. You can't add both or subtract both.

Original Numbers: 29·73
Step 1: $29 = 25 + 4 = 5^2 + 2^2$; $73 = 64 + 9 = 8^2 + 3^2$; $\{5, 2\}, \{8, 3\}$
Step 2: $5 \cdot 8 = 40$; $2 \cdot 3 = 6$
Step 3: $5 \cdot 3 = 15$; $2 \cdot 8 = 16$
Step 4:
$$\begin{cases} (40 + 6)^2 + (16 - 15)^2 = 46^2 + 1^2 = 2{,}116 + 1 = 2{,}117 \\ \qquad\qquad\qquad or \\ (40 - 6)^2 + (16 + 15)^2 = 34^2 + 31^2 = 1{,}156 + 961 = 2{,}117 \end{cases}$$

Original Numbers: 37·89
Step 1: $37 = 36 + 1 = 6^2 + 1^2$; $89 = 64 + 25 = 8^2 + 5^2$; $\{6, 1\}, \{8, 5\}$
Step 2: $6 \cdot 8 = 48$; $1 \cdot 5 = 5$
Step 3: $6 \cdot 5 = 30$; $1 \cdot 8 = 8$
Step 4:
$$\begin{cases} (48 + 5)^2 + (30 - 8)^2 = 53^2 + 22^2 = 2{,}809 + 484 = 3{,}293 \\ \qquad\qquad\qquad or \\ (48 - 5)^2 + (30 + 8)^2 = 43^2 + 38^2 = 1{,}849 + 1{,}444 = 3{,}293 \end{cases}$$

This method is usually difficult to use, so it's best to use it only when there is no other method that would be easy to use. This method can be useful in the following examples.

1.   Result involves the square of a number close to a round number.

Original Numbers: 173·281
Step 1: $173 = 169 + 4 = 13^2 + 2^2$; $281 = 256 + 25 = 16^2 + 5^2$; $\{13, 2\}, \{16, 5\}$
Step 2: $13 \cdot 16 = 208$; $2 \cdot 5 = 10$
Step 3: $13 \cdot 5 = 65$; $2 \cdot 16 = 32$

Your choices are between $218^2 + 33^2$ and $198^2 + 97^2$. $198^2 + 97^2$ may be easier to do because both are close to a round number. $39{,}204 + 9{,}409 = 48{,}613$.

2.   Result involves the square of a number divisible by 10.

Original Numbers: 137·53
Step 1: $137 = 121 + 16 = 11^2 + 4^2$; $53 = 49 + 4 = 7^2 + 2^2$; $\{11, 4\}, \{7, 2\}$
Step 2: $11 \cdot 7 = 77$; $4 \cdot 2 = 8$

Step 3: 11·2 = 22; 4·7 = 28

Your choices are between 85²+6² and 69²+50². 69²+50² may be easy to do because one of the squares ends in two zeros. 4,761+2,500=7,261

3.   Result involves the square of a small number.

Using the same example as before, 85²+6² may be easy to do because one of the numbers is small. 7,225+36=7,261

➢   **Sum Of Squares Method** (equation form)

$$x = a^2 + b^2$$

$$y = c^2 + d^2$$

$$xy = (a^2 + b^2)(c^2 + d^2) = (ac \pm bd)^2 + (ad \mp bc)^2$$

CHAPTER

STRETCH OUT METHOD

As you learned in Chapter 2, the easiest way to multiply two close numbers is the Stretch Out Method. This chapter will involve multiplying 2-digit or 3-digit numbers. You don't have to use the exact same estimates as used in the examples.

Step 1: Make the numbers easier to multiply by adding something to one number and subtracting the same thing from the other number. The resulting numbers are your estimates.
Step 2: Multiply the two estimates.
Step 3: Find the difference between either one original number and each estimate, or one estimate and each original number.
Step 4: Multiply the two differences.
Step 5: If the original numbers are closer to each other than the estimates, add Steps 2 and 4. If the original numbers are further away, subtract Steps 2 and 4.

Original Numbers: 73·66
Step 1: 73 + 2 = 75; 66 – 2 = 64
Step 2: 75·64 = 4,800
Step 3: 75 – 73 = 2 *or* 66 – 64 = 2; 75 – 66 = 9 *or* 73 – 64 = 9

Step 4: $2 \cdot 9 = 18$
Step 5: original numbers closer; $4{,}800 + 18 = 4{,}818$

Original Numbers: $47 \cdot 59$
Step 1: $47 + 3 = 50$; $59 - 3 = 56$
Step 2: $50 \cdot 56 = 2{,}800$
Step 3: $50 - 47 = 3$ *or* $59 - 56 = 3$; $59 - 50 = 9$ *or* $56 - 47 = 9$
Step 4: $3 \cdot 9 = 27$
Step 5: original numbers further; $2{,}800 - 27 = 2{,}773$

> ➤ **Stretch Out Method: Negative Stretch Out Variation**

Another option is to do the Negative Stretch Out Method. It's called the Negative Stretch Out Method because it's exactly the same as making one of the numbers negative then doing the Stretch Out Method.

Step 1: Make the numbers easier to multiply by adding or subtracting the same amount to both numbers. The resulting numbers are your estimates.
Step 2: Multiply the two estimates.
Step 3: Find the sum of an original number and the other number's estimate.
Step 4: Multiply that sum by the number you added or subtracted in Step 1.
Step 5: If you added in Step 1, subtract the results from Steps 2 and 4. If you subtracted in Step 1, add the results from Steps 2 and 4.

This is especially useful when both numbers have the same ones digit, because adding or subtracting to both can make them both divisible by 10.

Original Numbers: $43 \cdot 73$
Step 1: $43 - 3 = 40$; $73 - 3 = 70$
Step 2: $40 \cdot 70 = 2{,}800$
Step 3: $43 + 70$ *or* $40 + 73 = 113$
Step 4: $113 \cdot 3 = 339$
Step 5: $2{,}800 + 339 = 3{,}139$

Original Numbers: $38 \cdot 78$
Step 1: $38 + 2 = 40$; $78 + 2 = 80$
Step 2: $40 \cdot 80 = 3{,}200$
Step 3: $38 + 80$ *or* $40 + 78 = 118$
Step 4: $118 \cdot 2 = 236$
Step 5: $3{,}200 - 236 = 2{,}964$

➢ **Stretch Out Method: Single Estimate Variation**

The Single Estimate Method is very useful for multiplying two 3-digit numbers that are close together. It is just like using the Stretch Out Method but focusing on a single estimate.

Step 1: Pick a nearby number as an estimate.
Step 2: Subtract the estimate from each number. Sign matters. Your new numbers are called the "differences".
Step 3: Square the estimate.
Step 4: Add the two differences and multiply by the estimate.
Step 5: Multiply the differences.
Step 6: Add the results of Steps 3 through 5.

Your estimate for Step 1 can actually be any number, but it's easiest if you pick a nearby round number (multiple of 100 for three digit numbers). Here are some examples:

Original Numbers: 104·107
Step 1: 100
Step 2: 104 − 100 = 4; 107 − 100 = 7
Step 3: 10,000
Step 4: 100(4 + 7) = 1,100
Step 5: 4·7 = 28
Step 6: 10,000 + 1,100 + 28 = 11,128

Original Numbers: 294·313
Step 1: 300
Step 2: 294 − 300 =− 6; 313 − 300 = 13
Step 3: 90,000
Step 4: 300(− 6 + 13) = 2,100
Step 5: − 6·13 =− 78
Step 6: 90,000 + 2,100 − 78 = 92,022

If the numbers are both far away from a round number, you could estimate to halfway between round numbers. The next examples will show three ways to multiply 242·254, by estimating 200, 300, and 250.

Original Numbers: 242·254
Step 1: 200
Step 2: 242 − 200 = 42; 254 − 200 = 54

Step 3: 40,000
Step 4: 200(42 + 54) = 19,200
Step 5: 42·54 = 2,268
Step 6: 40,000 + 19,200 + 2,268 = 61,468

Original Numbers: 242·254
Step 1: 300
Step 2: 242 – 300 =– 58; 254 – 300 =– 46
Step 3: 90,000
Step 4: 300(– 58 +– 46) =– 31,200
Step 5: – 58· – 46 = 2,668
Step 6: 90,000 – 31,200 + 2,668 = 61,468

Original Numbers: 242·254
Step 1: 250
Step 2: 242 – 250 =– 8; 254 – 250 = 4
Step 3: 62,500
Step 4: 250(– 8 + 4) =– 1,000
Step 5: – 8·4 =– 32
Step 6: 62,500 – 1,000 – 32 = 61,468

With the estimate being 250 in the last example, even though Step 3 is harder because there are more non-zero numbers to remember, it may be better because Steps 4-6 are much easier.

> ➢ **Stretch Out Method: Single Estimate Variation** (alternate steps)

Here's a different way to do the Single Estimate Method.

Step 1: Pick a nearby number as an estimate.
Step 2: Add the two original numbers.
Step 3: Subtract your estimate from the result from Step 2, then multiply by your estimate.
Step 4: Add, to the result from Step 3, the product of the differences from your estimate. Sign matters.

Original Numbers: 104·107
Step 1: 100
Step 2: 104 + 107 = 211
Step 3: 100(211 – 100) = 11,100
Step 4: 11,100 + (104 – 100)(107 – 100) = 11,128

Original Numbers: 294·313
Step 1: 300
Step 2: 294 + 313 = 607
Step 3: 300(607 − 300) = 92,100
Step 4: 92,100 + (294 − 300)(313 − 300) = 92,022

Original Numbers: 242·254
Step 1: 250
Step 2: 242 + 254 = 496
Step 3: 250(496 − 250) = 61,500
Step 4: 61,500 + (242 − 250)(254 − 250) = 61,468

➤ **Stretch Out Method: Factoring First**

Since the Stretch Out Method is best when the numbers are close together, factoring can be done to make them closer. If one of the numbers is close to double the other, you could either double the lower number or (if even) halve the larger number, then do the opposite at the end. In the first example, 34 is doubled to 68 so it's closer to 63. In the second example, 78 is halved to 39 so it's closer to 47.

$$63 \cdot 34 = \frac{63 \cdot 68}{2} = \frac{61 \cdot 70 + 2 \cdot 7}{2} = \frac{4{,}270 + 14}{2} = \frac{4{,}284}{2} = 2{,}142$$

$$78 \cdot 47 = 2 \cdot 39 \cdot 47 = 2(43^2 - 4^2) = 2(1{,}849 - 16) = 2 \cdot 1{,}833 = 3{,}666$$

Another option is to figure out the prime factors, then form two numbers that are close together. For example, if you're multiplying 28·69, you could double the 28 to get 56.

$$28 \cdot 69 = \frac{56 \cdot 69}{2} = \frac{60 \cdot 65 - 4 \cdot 9}{2} = \frac{3{,}900 - 36}{2} = \frac{3{,}864}{2} = 1{,}932$$

But if you separated 28·69 into their prime factors, you get 2·2·7·3·23. The closest two numbers to each other that you can get from multiplying those are 42·46. Since those are very close and both even, you can use the Difference of Squares Method from Chapter 5.

$$42 \cdot 46 = 44^2 - 2^2 = 1{,}936 - 4 = 1{,}932$$

As long as you're good with squares, 42·46 (by finding the prime factors) is much easier to do than 56·69/2 (by doubling one of the numbers).

> ➢ **Stretch Out Method: Far Apart Numbers**

The Stretch Out Method and Single Estimate Method are best for multiplying numbers that are close together. When multiplying three-digit numbers, it's very likely that the numbers won't be close together. If that happens, besides using the Proportional Stretch Out Method (shown after these methods), there are three possible options you can use.

> ➢ **Stretch Out Method: Far Apart Numbers** (Option 1)

Use the Stretch Out Method (or Single Estimate Variation), keeping the big difference.

Step 1: Pick a nearby number as an estimate.
Step 2: Subtract the estimate from each number. Sign matters. Your new numbers are called the "differences".
Step 3: Square the estimate.
Step 4: Add the two differences and multiply by the estimate.
Step 5: Multiply the differences.
Step 6: Add the results of Steps 3 through 5.

Original Numbers: 332·749
Step 1: 300
Step 2: 332 − 300 = 32; 749 − 300 = 449
Step 3: $300^2 = 90,000$
Step 4: 300(32 + 449) = 144,300
Step 5: 32·449 = 14,368
Step 6: 90,000 + 144,300 + 14,368 = 248,668

That option is sometimes useful, but it's not the best in this case because Step 5 involves multiplying 32·449 and Step 6 involves adding many significant digits. Luckily, there are other options.

> ➢ **Stretch Out Method: Far Apart Numbers** (Option 2)

Double or halve one of the numbers to make them closer to each other, then undo that at the end. In this example, you multiply 332 by 2 to make it closer to 749, so you'll divide your solution by 2 at the end.

$Original\ Numbers$: 332·749
$Factored\ Numbers$: 664·749
$Step\ 1$: 700
$Step\ 2$: 664 − 700 =− 36; 749 − 700 = 49
$Step\ 3$: $700^2 = 490{,}000$
$Step\ 4$: 700(− 36 + 49) = 9,100
$Step\ 5$: − 36·49 =− 1,764
$Step\ 6$: 490,000 + 9,100 − 1,764 = 497,336
$Undo\ Factoring$: $\dfrac{497{,}336}{2} = 248{,}668$

> ➤ **Stretch Out Method: Far Apart Numbers** (Option 3)

Subtract from the larger number's hundreds place to make the numbers closer, then undo that at the end. In this example, you subtracted 400 from 749 to make it closer to 332. Because you're only doing 349·332 instead of 749·332, you'll add 400·332 at the end.

$Original\ Numbers$: 332·749
$Distributed\ Numbers$: 332·349 + 332·400
$Step\ 1$: 300
$Step\ 2$: 332 − 300 = 32; 349 − 300 = 49
$Step\ 3$: $300^2 = 90{,}000$
$Step\ 4$: 300(32 + 49) = 24,300
$Step\ 5$: 32·49 = 1,568
$Step\ 6$: 90,000 + 24,300 + 1,568 = 115,868
$Add\ Distributed\ Part$: 115,868 + 332·400
  = 115,868 + 132,800 = 248,668

> ➤ **Proportional Stretch Out Method**

The regular Stretch Out Method works best for multiplying close numbers, because if they're far apart, it's harder to figure out the difference between the low number and the high estimate or between the high number and the low estimate. If you use the Proportional Stretch Out Method, you won't have to do that.

Step 1: Estimate a ratio comparing the higher number to the lower number. You can actually pick any random ratio for this, but the other steps will be easier if the ratio is a close estimate.

Step 2: Add something to the smaller number and subtract a proportional amount (following the ratio in Step 1) from the larger number to get numbers that are easy to multiply in your head (you could also subtract from the smaller and add to the larger). It's best to change the numbers as little as possible, to make Step 5 easier. The two resulting numbers are the Stretch Out Estimates.

Step 3: Find nearby numbers that have the exact ratio as Step 1. You can use any numbers that have that ratio, but it's best if one of the numbers matches an original number or a Stretch Out Estimate, because that will allow you to skip either Step 4 or 5. If you can't match exactly, make it as close as possible. The two resulting numbers are the Proportional Estimates.

Step 4: Multiply the differences between the Proportional Estimate numbers and the original numbers. If the differences are in the same direction, the result of this step is positive. If the differences are towards each other or away from each other, the result of this step is negative. If either of the Proportional Estimate numbers is equal to the corresponding original number, you can skip this step.

Step 5: Multiply the differences between the Proportional Estimate numbers and the Stretch Out Estimate numbers. If the differences are in the same direction, the result of this step is negative. If the differences are towards each other or away from each other, the result of this step is positive. (Note that the previous two sentences are the opposite of Step 4). If either of the Proportional Estimate numbers is equal to the corresponding Stretch Out Estimate number, you can skip this step.

Step 6: Multiply the Stretch Out Estimate numbers.

Step 7: Add the results of Steps 4-6.

This method requires a lot of memory. To prevent losing your place, it's most likely easier to add the results of Steps 4 and 5 before doing Step 6.

The examples used for this method are the examples that have already been shown using other methods. While only one example is shown for the estimates and ratios, you could use other estimates or ratios.

Original Numbers: $7 \cdot 16$
Step 1: $2:1$
Step 2: $7 + 3 = 10$; $16 - 3 \cdot 2 = 10$; $(10, 10)$

Step 3: (7, 14)

Step 4: 0 (*one number matches*)

Step 5: $10 - 7 = 3$; $14 - 10 = 4$; *toward or away*; $3 \cdot 4 = 12$

Step 6: $10 \cdot 10 = 100$

Step 7: $100 + 12 + 0 = 112$

Original Numbers: $34 \cdot 63$

Step 1: 2:1

Step 2: $34 - 4 = 30$; $63 + 4 \cdot 2 = 71$; (30, 71)

Step 3: (34, 68)

Step 4: 0 (*one number matches*)

Step 5: $34 - 30 = 4$; $71 - 68 = 3$; *toward or away*; $4 \cdot 3 = 12$

Step 6: $30 \cdot 71 = 2,130$

Step 7: $2,130 + 12 + 0 = 2,142$

Original Numbers: $47 \cdot 78$

Step 1: 5:3

Step 2: $47 + 3 = 50$; $78 - 3 \cdot \dfrac{5}{3} = 73$; (50, 73)

Step 3: (48, 80)

Step 4: $48 - 47 = 1$; $80 - 78 = 2$; *same direction*; $1 \cdot 2 = 2$

Step 5: $50 - 48 = 2$; $80 - 73 = 7$; *toward or away*; $2 \cdot 7 = 14$

Step 6: $50 \cdot 73 = 3,650$

Step 7: $3,650 + 14 + 2 = 3,666$

Original Numbers: $73 \cdot 449$

Step 1: 6:1

Step 2: $73 + 2 = 75$; $449 - 2 \cdot 6 = 437$; (75, 437)

Step 3: (75, 450)

Step 4: $75 - 73 = 2$; $450 - 449 = 1$; *same direction*; $1 \cdot 2 = 2$

Step 5: 0 (*one number matches*)

Step 6: $75 \cdot 437 = 32,775$

Step 7: $32,775 + 2 = 32,777$

Original Numbers: $332 \cdot 749$

Step 1: 5:2

Step 2: $332 - 32 = 300$; $749 + 32 \cdot \dfrac{5}{2} = 829$; (300, 829)

Step 3: (300, 750)

Step 4: $332 - 300 = 32$; $750 - 749 = 1$; *toward or away*; $-32 \cdot 1 = -32$

Step 5: 0 (*one number matches*)

Step 6: $300 \cdot 829 = 248,700$

Step 7: 248,700 − 32 + 0 = 248,668

> ➤ **Proportional Stretch Out Method** (finding stretch out estimate before finding ratio)

If you want to make sure your Stretch Out Estimates will be easy to multiply, you can first figure out how much you want to add to or subtract from one of the numbers, and make sure the ratio includes that number. For example, when multiplying 263·442, it would be great if you could subtract 13 from 263 to get 250, because it's easier to multiply by 250 than by something else nearby. Because of that, the ratio should be x:13. Since 263/13 is close to 20, x should be close to 442/20.

Original Numbers: 263·442
Step 1: 22:13
Step 2: 263 − 13 = 250; 442 + 22 = 464; (250, 464)
Step 3: (260, 440)
Step 4: 263 − 260 = 3; 442 − 440 = 2; *same direction*; 3·2 = 6
Step 5: 260 − 250 = 10; 464 − 440 = 24; *toward or away*; 10·24 = 240
Step 6: 250·464 = 116,000
Step 7: 116,000 + 240 + 6 = 116,246

For the next one, 742 would be best stretched out to 750, so this would be easiest with a ratio of 8:x.

Original Numbers: 473·742
Step 1: 8:5
Step 2: 473 − 5 = 468; 742 + 8 = 750; (468, 750)
Step 3: (475, 760)
Step 4: 475 − 473 = 2; 760 − 742 = 18; *same direction*; 2·18 = 36
Step 5: 475 − 468 = 7; 760 − 750 = 10; *same direction*; − 7·10 =− 70
Step 6: 468·750 = 351,000
Step 7: 351,000 − 70 + 36 = 350,966

> ➤ **Proportional Stretch Out Method** (fractional stretch out estimate)

In the previous examples, you were able to choose one of the Stretch Out Estimates before choosing the ratio, but you still had to do both Steps 4 and 5. If you want to skip Step 5 by making a Stretch Out Estimate match a Proportional Estimate, you can make the other Stretch Out Estimate not a whole number.

That may seem difficult, but your result from multiplying both Stretch Out Estimates will be a whole number even though one of the estimates is not a whole number.

The same two examples are being used as before. Since one of the Stretch Out Estimates is not a whole number, you should multiply the integer part and fraction part separately as shown in Step 6.

Original Numbers: 263·442
Step 1: 9:5
Step 2: $263 - 13 = 250; 442 + 13 \cdot \frac{9}{5} = 465\frac{2}{5}; (250, 465\frac{2}{5})$

Step 3: (250, 450)
Step 4: $263 - 250 = 13; 450 - 442 = 8;$ *toward or away;* $-13 \cdot 8 = -104$
Step 5: 0 *(one number matches)*
Step 6: $250 \cdot 465\frac{2}{5} = 250 \cdot 465 + 250 \cdot \frac{2}{5} = 116{,}250 + 100 = 116{,}350$

Step 7: $116{,}350 - 104 = 116{,}246$

Original Numbers: 473·742
Step 1: 3:2
Step 2: $473 - 8 \cdot \frac{2}{3} = 467\frac{2}{3}; 742 + 8 = 750; \left(467\frac{2}{3}, 750\right)$

Step 3: (500, 750)
Step 4: $500 - 473 = 27; 750 - 742 = 8;$ *same direction;* $27 \cdot 8 = 216$
Step 5: 0 *(one number matches)*
Step 6: $467\frac{2}{3} \cdot 750 = 467 \cdot 750 + \frac{2}{3} \cdot 750 = 350{,}250 + 500 = 350{,}750$

Step 7: $350{,}750 + 216 = 350{,}966$

➢ **Proportional Stretch Out Method** (tens and ones place divisible by hundreds place)

The Proportional Stretch Out Method is usually the easiest to use whenever one of the numbers has a combined tens and ones place that's divisible by the hundreds place. This is because you can use that hundreds place as both the Proportional Estimate and the Stretch Out Estimate.

You can round the hundreds place up and make the tens and ones place negative. For example, if one of the numbers is 679, 79 is not divisible by 6, but -21 is divisible by 7.

Original Numbers: 428·739
Step 1: 7:4

Step 2: $428 - 28 = 400; 739 + 28 \cdot \dfrac{7}{4} = 788; (400, 788)$

Step 3: $(400, 700)$
Step 4: $428 - 400 = 28; 739 - 700 = 39; same\ direction; 28 \cdot 39 = 1,092$
Step 5: $0\ (one\ number\ matches)$
Step 6: $400 \cdot 788 = 315,200$
Step 7: $315,200 + 1,092 = 316,292$

> ➢ **Proportional Stretch Out Method** (factor first)

Since one of the numbers has a tens and ones place divisible by the hundreds place, you can divide that number by the hundreds place then multiply the other number by what you divided the first number by.

Original Numbers: 428·739

Factored Numbers: $\dfrac{428}{4} \cdot 739 \cdot 4 = 107 \cdot 2,956$

Step 1: 30:1
Step 2: $107 - 7 = 100; 2,956 + 7 \cdot 30 = 3,166; (100, 3166)$
Step 3: $(100, 3000)$
Step 4: $107 - 100 = 7; 3,000 - 2,956 = 44; toward\ or\ away; -7 \cdot 44 = -308$
Step 5: $0\ (one\ number\ matches)$
Step 6: $100 \cdot 3,166 = 316,600$
Step 7: $316,600 - 308 = 316,292$

> ➢ **Proportional Stretch Out Method** (tens and ones place not divisible by hundreds place)

If the tens and ones place is not divisible by the hundreds place, you have two options. One is to pretend it's divisible and use non-integer values. Examples are shown for not factoring first and for factoring first.

Original Numbers: 319·537
Step 1: 5:3

Step 2: $319 - 19 = 300; 537 + 19 \cdot \dfrac{5}{3} = 568\dfrac{2}{3}; \left(300; 568\dfrac{2}{3}\right)$

Step 3: (300, 500)
Step 4: 319 − 300 = 19; 537 − 500 = 37; *same direction*; 19·37 = 703
Step 5: 0 (*one number matches*)
Step 6: 300·568$\frac{2}{3}$ = 170,600

Step 7: 170,600 + 703 = 171,303

Original Numbers: 319·537
Factored Numbers: $\frac{319}{3}$·537·3 = 106$\frac{1}{3}$·1,611

Step 1: 16:1
Step 2: 106$\frac{1}{3}$ − 6$\frac{1}{3}$ = 100; 1,611 + 6$\frac{1}{3}$·16 = 1,712$\frac{1}{3}$; $\left(100, 1712\frac{1}{3}\right)$

Step 3: (100, 1600)
Step 4: 106$\frac{1}{3}$ − 100 = 6$\frac{1}{3}$; 1,611 − 1,600 = 11; *same direction*;

6$\frac{1}{3}$·11 = 69$\frac{2}{3}$

Step 5: 0 (*one number matches*)
Step 6: 100·1,712$\frac{1}{3}$ = 171,233$\frac{1}{3}$

Step 7: 171,233$\frac{1}{3}$ + 69$\frac{2}{3}$ = 171,303

> ➢ **Proportional Stretch Out Method** (distribute to make tens and ones place divisible by hundreds place)

Instead of using non-integer values, you could distribute first.

Original Numbers: 319·537
Distributed Numbers: 318·537 + 537
Step 1: 5:3
Step 2: 318 − 18 = 300; 537 + 18·$\frac{5}{3}$ = 567; (300; 567)

Step 3: (300, 500)
Step 4: 318 − 300 = 18; 537 − 500 = 37; *same direction*; 18·37 = 666
Step 5: 0 (*one number matches*)
Step 6: 300·567 = 170,100
Step 7: 170,100 + 666 = 170,766
Add Distributed Part: 170,766 + 537 = 171,303

Original Numbers: 319·537

Distributed Numbers: 318·537 + 537

Factored Numbers: $\dfrac{318}{3}$·537·3 + 537 = 106·1,611 + 537

Step 1: 16:1
Step 2: 106 − 6 = 100; 1,611 + 6·16 = 1,707; (100, 1707)
Step 3: (100, 1600)
Step 4: 106 − 100 = 6; 1,611 − 1,600 = 11; *same direction*; 6·11 = 66
Step 5: 0 (*one number matches*)
Step 6: 100·1,707 = 170,700
Step 7: 170,700 + 66 = 170,766
Add Distributed Part: 170,766 + 537 = 171,303

> ➢ **Proportional Stretch Out Method** (equation form)

Original Numbers: (x, y)
Proportional Estimates: (a, ka)
Stretch Out Estimates: $(x - n, y + kn)$
$xy = (x - n)(y + kn) + (x - a)(y - ka) - (x - n - a)(y + kn - ka)$

> ➢ **Proportional Stretch Out Method** (number line illustration)

The Proportional Stretch Out Method can also be explained using number lines.

These are number line illustrations for the examples shown before. The big dots represent your original numbers. The stars represent your Stretch Out Estimates. The vertical line represents your Proportional Estimates. You'll notice that the distances between the Stretch Out Estimates and your original numbers are proportional to each other, meaning that by looking at the scaled diagram, they should appear to be the same distance.

332·749

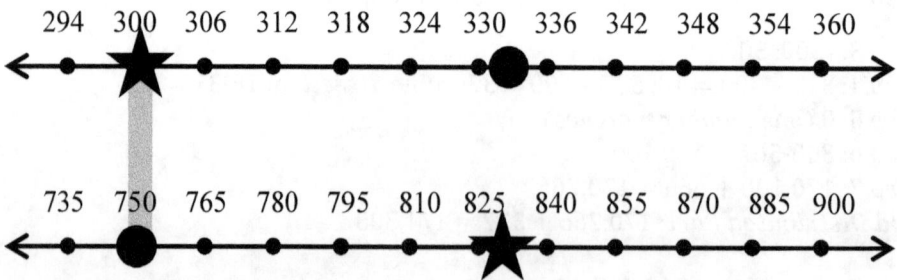

263·442

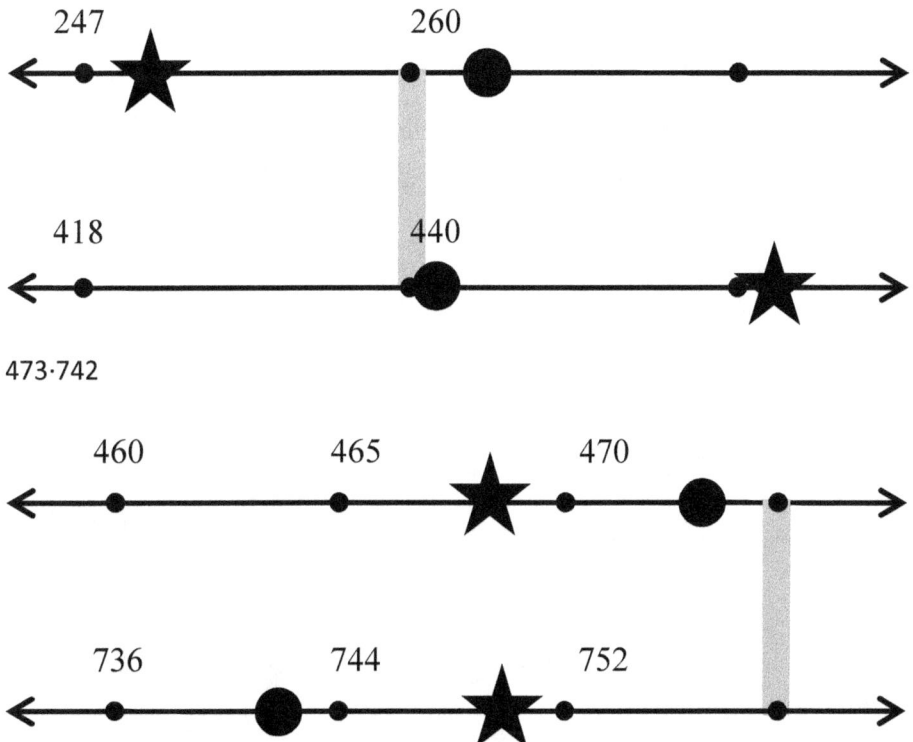

473·742

All of these diagrams can be safely altered by doing any of the following:

- moving the vertical line
- moving a star, and moving the other star a proportional amount in the other direction
- using an entirely different ratio between the two number lines

After creating the diagram, your solution is the sum of these three steps.

Step 1: Multiply the positions of the stars.

Step 2: If one of the big dots is on the vertical line, skip this step. If the big dots are on the same side of the vertical line, multiply the distances from the vertical line and make it positive. If the big dots are on opposite sides of the vertical line, multiply the distances from the vertical line and make it negative.

Step 3: If one of the stars is on the vertical line, skip this step. If the stars are on the same side of the vertical line, multiply the distances from the vertical line and make it negative. If the stars are on opposite sides of the vertical line, multiply the distances from the vertical line and make it positive.

The regular Stretch Out Method is when the ratio between the two number lines is 1:1 and when one of the stars or dots is on the vertical line.

CHAPTER

ALMOST-ROUND MULTIPLE METHOD

The Almost-Round Multiple Method is a type of factoring that's used to make one of the numbers close to a multiple of 100 so it's easier to multiply. Here are the steps.

> **Step 1:** Find a multiple of the first number (you can switch which number is first) that's close to a multiple of 100. Keep track of the multiple and what number you multiplied by to get that multiple.
> **Step 2:** Divide the second number by what you multiplied the first number by, and keep track of the whole part and the remainder.
> **Step 3:** Multiply the remainder by the first number.
> **Step 4:** Multiply the whole part by the almost-round multiple, and add to the result from Step 3.

For Step 1, the following table is a list of almost-round multiples for all numbers 6 through 94. All are the first multiple that is 4 or less away from a multiple of 100, unless the original number is divisible by 5 (because the closest possible is 5 away), or is 37 or 74 (because 111 and 222 are easy to multiply using the Standard Method). The number in parentheses is what you multiply by to get the multiple. Besides 37 and 74, the multiplier used to get the almost round multiple for one number is the same as it is for 100 minus that number (for example, 23 and 77 both multiply by 13, 38 and 62 both multiply by 8, etc).

#	Multiple	#	Multiple	#	Multiple	#	Multiple	#	Multiple	#	Multiple
6	102 (17)	21	399 (19)	36	396 (11)	51	102 (2)	66	198 (3)	81	1,296 (16)
7	98 (14)	22	198 (9)	37	111(3)	52	104 (2)	67	201 (3)	82	902 (11)
8	96 (12)	23	299 (13)	38	304 (8)	53	901 (17)	68	204 (3)	83	498 (6)
9	99 (11)	24	96 (4)	39	702 (18)	54	702 (13)	69	897 (13)	84	504 (6)
10	100 (10)	25	100 (4)	40	200 (5)	55	495 (9)	70	700 (10)	85	595 (7)
11	99 (9)	26	104 (4)	41	697 (17)	56	504 (9)	71	497 (7)	86	602 (7)
12	96 (8)	27	297 (11)	42	504 (12)	57	399 (7)	72	504 (7)	87	696 (8)
13	104 (8)	28	196 (7)	43	301 (7)	58	696 (12)	73	803 (11)	88	704 (8)
14	98 (7)	29	203 (7)	44	396 (9)	59	1,003 (17)	74	222 (3)	89	801 (9)
15	105 (7)	30	300 (10)	45	405 (9)	60	300 (5)	75	300 (4)	90	900 (10)
16	96 (6)	31	403 (13)	46	598 (13)	61	1,098 (18)	76	304 (4)	91	1,001 (11)
17	102 (6)	32	96 (3)	47	799 (17)	62	496 (8)	77	1,001 (13)	92	1,104 (12)
18	198 (11)	33	99 (3)	48	96 (2)	63	504 (8)	78	702 (9)	93	1,302 (14)
19	304 (16)	34	102 (3)	49	98 (2)	64	704 (11)	79	1,501 (19)	94	1,598 (17)
20	100 (5)	35	105 (3)	50	100 (2)	65	195 (3)	80	400 (5)		

Here are some examples. If the multiple used is slightly less than a multiple of 100 instead of slightly more, it is instead shown as the multiple of 100 minus the difference (497 is shown as 500-3).

Original Numbers: 17·254
Step 1: 17·6 = 102
Step 2: $\dfrac{254}{6} = 42\dfrac{2}{6}$
Step 3: 2·17 = 34
Step 4: 42·102 = 4,284; 4,284 + 34 = 4,318

Original Numbers: 67·218
Step 1: 67·3 = 201
Step 2: $\dfrac{218}{3} = 72\dfrac{2}{3}$
Step 3: 2·67 = 134
Step 4: 72·201 = 14,472; 14,472 + 134 = 14,606

Original Numbers: 71·187

Step 1: 71·7 = 497 = 500 − 3

Step 2: $\dfrac{187}{7} = 26\dfrac{5}{7}$

Step 3: 5·71 = 355

Step 4: 26·(500 − 3) = 13,000 − 78 = 12,922; 12,922 + 355 = 13,277

For three digit numbers, use the same almost-round multiple as you would for its last two digits. It's usually hard to do if there's a remainder (like the three previous examples), so it's best to use only if there's no remainder. You can distribute the almost-round multiple by separating the hundreds place and the ones place (2,502 becomes 2,500+2).

Original Numbers: 417·672

Step 1: 417·6 = 2,502

Step 2: $\dfrac{672}{6} = 112$

Step 3: 0 *(no remainder)*

Step 4: 112·2,502 = 112(2,500 + 2) = 280,000 + 224 = 280,224

Original Numbers: 567·423

Step 1: 567·3 = 1,701

Step 2: $\dfrac{423}{3} = 141$

Step 3: 0 *(no remainder)*

Step 4: 141·1,701 = 141(1,700 + 1) = 239,700 + 141 = 239,841

Original Numbers: 743·238

Step 1: 743·7 = 5,201

Step 2: $\dfrac{238}{7} = 34$

Step 3: 0 *(no remainder)*

Step 4: 34·5,201 = 34(5,200 + 1) = 176,800 + 34 = 176,834

Because it's nearly impossible to memorize that entire table, it's best to only memorize a few and use a different method for other numbers. You can also use the following tips if you can't memorize the whole table.

1. Only memorize the primes. If a number is not prime, you can just factor it so it becomes prime.

2. Remember common decimals. Since .333 and .666 are decimals created by dividing by 3, numbers like 33 and 67 can be multiplied by 3 to get an almost-round multiple. Since .142, .285, .428, .571, .714, and .857 are decimals created by dividing by 7, numbers like 14, 29, 43, 57, 71, and 86 can be multiplied by 7 to get an almost-round multiple.

CHAPTER

3-WAY MULTIPLICATION

This chapter involves multiplication of three two-digit numbers.

> **Three-Way Single Estimate Method** (for close numbers)

If all three numbers are close together, the easiest method is usually the Three-Way Single Estimate Method.

Step 1: Choose a number to estimate. You can technically pick any number, but it's best if you pick a multiple of 10 that's close to your original numbers.

Step 2: Find the difference between the estimate and each of the original numbers (sign matters, original minus estimate).

Step 3: Add the three differences.

Step 4: Add the products of the first and second differences, first and third differences, and second and third differences.

Step 5: Multiply the three differences.

Step 6: Start with the estimate. Add the result from Step 3. Multiply by the estimate. Add the result from Step 4. Multiply by the estimate. Add the result from Step 5.

Here are some examples.

Original Numbers: 12·13·17
Step 1: 10
Step 2: 2, 3, 7
Step 3: $2 + 3 + 7 = 12$
Step 4: $2 \cdot 3 + 2 \cdot 7 + 3 \cdot 7 = 41$
Step 5: $2 \cdot 3 \cdot 7 = 42$
Step 6: $10 + 12 = 22$; $22 \cdot 10 = 220$; $220 + 41 = 261$;
$261 \cdot 10 = 2{,}610$; $2{,}610 + 42 = 2{,}652$

Original Numbers: 19·24·31
Step 1: 20
Step 2: $-1, 4, 11$
Step 3: $-1 + 4 + 11 = 14$
Step 4: $-1 \cdot 4 + -1 \cdot 11 + 4 \cdot 11 = 29$
Step 5: $-1 \cdot 4 \cdot 11 = -44$
Step 6: $20 + 14 = 34$; $34 \cdot 20 = 680$; $680 + 29 = 709$;
$709 \cdot 20 = 14{,}180$; $14{,}180 - 44 = 14{,}136$

For Step 4, you could make it easier by adding any two of the differences, multiplying by the other difference, and adding the product of the two differences you added before. In the previous examples, when your differences were 2, 3, and 7, you could do (2+3)·7+2·3, (2+7)·3+2·7, or (3+7)·2+3·7. When the differences were -1, 4, and 11, you could do (-1+4)·11+-1·4, (-1+11)·4+-1·11, or (4+11)·-1+4·11.

➢ **Three-Way Single Estimate Method** (using single digit estimate)

Since the estimate can be anything, it may be useful to make the estimate a single digit, especially if all three numbers have the same ones digit.

Original Numbers: 23·33·53
Step 1: 3
Step 2: 20, 30, 50
Step 3: $20 + 30 + 50 = 100$
Step 4: $20 \cdot 30 + 20 \cdot 50 + 30 \cdot 50 = 3{,}100$
Step 5: $20 \cdot 30 \cdot 50 = 30{,}000$
Step 6: $3 + 100 = 103$; $103 \cdot 3 = 309$; $309 + 3{,}100 = 3{,}409$;

$3,409 \cdot 3 = 10,227; 10,227 + 30,000 = 40,227$

If you're using a single digit for the estimate, it may be useful to make the estimate negative, because multiplying by -1, -2, or -3 may be easier than multiplying by 7, 8, or 9.

Original Numbers: 38·48·78
Step 1: – 2
Step 2: 40, 50, 80
Step 3: $40 + 50 + 80 = 170$
Step 4: $40 \cdot 50 + 40 \cdot 80 + 50 \cdot 80 = 9,200$
Step 5: $40 \cdot 50 \cdot 80 = 160,000$
Step 6: $-2 + 170 = 168; 168 \cdot -2 =- 336; -336 + 9,200 = 8,864;$
$8,864 \cdot -2 =- 17,728; -17,728 + 160,000 = 142,272$

If two of the ones digits are the same and the third ones digit is the opposite (10 minus it), pretend that the number with the opposite ones digit is negative. Your result will be negative because you made one of the original numbers negative, so just make the result positive at the end. In this example, you'll pretend the 57 is -57.

Original Numbers: 43·57·83
Step 1: 3
Step 2: 40, – 60, 80
Step 3: $40 – 60 + 80 = 60$
Step 4: $40 \cdot -60 + 40 \cdot 80 + 80 \cdot -60 =- 4,000$
Step 5: $40 \cdot -60 \cdot 80 =- 192,000$
Step 6: $3 + 60 = 63; 63 \cdot 3 = 189; 189 – 4,000 =- 3,811;$
$-3,811 \cdot 3 =- 11,433; -11,433 – 192,000 =- 203,433$
Remove Negative Sign: 203,433

If two of the ones digits are the same and the third ones digit is off by 5, it may still be possible to do in your head with one of the differences ending in 5.

Original Numbers: 32·42·47
Step 1: 2
Step 2: 30, 40, 45
Step 3: $30 + 40 + 45 = 115$
Step 4: $30 \cdot 40 + 30 \cdot 45 + 40 \cdot 45 = 4,350$
Step 5: $30 \cdot 40 \cdot 45 = 54,000$

Step 6: $2 + 115 = 117$; $117 \cdot 2 = 234$; $234 + 4{,}350 = 4{,}584$;
$4{,}584 \cdot 2 = 9{,}168$; $9{,}168 + 54{,}000 = 63{,}168$

➢ **Three-Way Single Estimate Method** (for any three numbers)

You now know how to multiply three numbers if they all end in the same ones digit, the opposite ones digit, or a ones digit off by 5. You can now multiply three numbers if they're all one away from a multiple of 5 (ends in 1, 4, 6, 9) or if they're all two away from a multiple of 5 (ends in 2, 3, 7, 8). This is because they'll always either add or subtract to a multiple of 5. If you're multiplying two numbers from one set and one number from the other set, you can make them all the same set by doubling or halving a number to switch it to the other set. If you do that, you must undo it at the end.

In this example, two of the numbers (42, 68) are two away from a multiple of 5, and one of the numbers (74) is one away. To change the one-away number to a two-away number, you can halve it (you can also double it, but since it's even it's easier to halve it). Since you halved one of the numbers, you'll have to double the result at the end.

Original Numbers: $42 \cdot 68 \cdot 74$
Adjustment: $42 \cdot 68 \cdot 37$
Step 1: 2
Step 2: $40, \ -70, 35$
Step 3: $40 - 70 + 35 = 5$
Step 4: $40 \cdot -70 + 40 \cdot 35 + 35 \cdot -70 =- 3{,}850$
Step 5: $40 \cdot -70 \cdot 35 =- 98{,}000$
Step 6: $2 + 5 = 7$; $7 \cdot 2 = 14$; $14 - 3{,}850 =- 3{,}836$;
$-3{,}836 \cdot 2 =- 7{,}672$; $-7{,}672 - 98{,}000 =- 105{,}672$
Undo Adjustment: $-105{,}672 \cdot -2 = 211{,}344$

➢ **Cubes** (Three-Way Single Estimate Method, altered)

If you're multiplying three of the same number, you can do a slightly altered version of the Three-Way Single Estimate Method. The original version still works, but this one is quicker for cubes. To do this, it is best to have the cubes of all numbers 1 through 9 memorized.

Step 1: Choose a number to estimate. You can technically pick any number, but it's best if you pick a multiple of 10 that's close to your original number.
Step 2: Find the difference between the estimate and the original number (sign matters, original minus estimate).
Step 3: Multiply the following: the number 3, the original number, the estimate, the original number minus the estimate.
Step 4: Add the cube of the estimate, the cube of the difference, and the result of Step 3.

Original Number: 32
Step 1: 30
Step 2: 2
Step 3: $3 \cdot 32 \cdot 30 \cdot 2 = 5{,}760$
Step 4: $30^3 = 27{,}000; 2^3 = 8; 27{,}000 + 5{,}760 + 8 = 32{,}768$

Original Number: 67
Step 1: 70
Step 2: -3
Step 3: $3 \cdot 67 \cdot 70 \cdot -3 = -42{,}210$
Step 4: $70^3 = 343{,}000; -3^3 = -27; 343{,}000 - 42{,}210 - 27 = 300{,}763$

➢ **Cubes** (Consecutive Number Method)

Another option for cubing a number is the Consecutive Number Method, which involves changing the numbers into three consecutive numbers instead of three of the same number.

Step 1: Add something to one of the numbers and subtract that from another one of the numbers.
Step 2: Multiply the three new numbers (the original, the one you added to, the one you subtracted from) using any of the methods in previous chapters.
Step 3: Add the original number times the square of what you added and subtracted in Step 1.

For Step 2, both of the following examples can be solved by factoring. For $30 \cdot 32 \cdot 34$, you can triple the 34 to get 102 then divide the 30 by 3 to get 10, so your result will be $10 \cdot 32 \cdot 102$. For $66 \cdot 67 \cdot 68$, you can triple the 67 to get 201 and divide the 66 by 3 to get 22; you can then multiply the 68 by 2 to get 136 and divide the 22 by 2 to get 11, so your result will be $11 \cdot 201 \cdot 136$.

Original Number: 32
Step 1: 30, 32, 34
Step 2: 30·32·34 = 10·32·102 = 10·3,264 = 32,640
Step 3: 2^2 = 4; 32·4 = 128; 32,640 + 128 = 32,768

Original Number: 67
Step 1: 66, 67, 68
Step 2: 66·67·68 = 11·201·136 = 11·27,336 = [2][9][10][6][9][6]
= 300,696
Step 3: 1^2 = 1; 67·1 = 67; 300,696 + 67 = 300,763

➢ **Alternatives To Three-Way Single Estimate Method**

If the Three-Way Single Estimate method is too hard for the problem you're doing, you could use the following methods instead.

➢ **Standard Method**

If one of the three numbers is easy to multiply using the Standard Method, multiply the other two numbers first. For example, if you're multiplying 43·62·79, 79 is easy to multiply with the Standard Method because it's [1][-2][-1]. You can first multiply 43·62 using the Stretch Out Method, then multiply the result by 79 using the Standard Method.

43·62 = 40·65 + 3·22 = 2,600 + 66 = 2,666

2,666·[1][– 2][– 1] = [2][6 – 4][6 – 12 – 2][6 – 12 – 6][– 12 – 6][– 6]
= [2][2][– 8][– 12][– 18][– 6] = 210,614

➢ **Almost-Round Multiple Method**

If one of the three numbers is easy to multiply using the Almost-Round Multiple Method, multiply the other two numbers first. For example, if you're multiplying 53·67·82, 67 is easy to multiply because you can divide by 3 then multiply by 201. First multiply 53·82 using the Stretch Out Method, then multiply the result by 67 using the Almost-Round Multiple Method.

53·82 = 55·80 – 2·27 = 4,400 – 54 = 4,346

4,346·67 = 1,449·201 – 67 = 289,800 + 1,449 – 67 = 291,182

66 How To Be A Math Genius

➢ Two Numbers Multiply To Almost-Round Number

If two of the three numbers multiply to something close to a round number (tens or hundreds place is 9 or 0), multiply those two numbers first. For example, if you're multiplying $42 \cdot 57 \cdot 73$, you can use the Stretch Out Method to solve $42 \cdot 57 = 2,394$, which is $2,400 - 6$. You can then do $2,400 \cdot 73 - 6 \cdot 73$.

$$42 \cdot 57 = 49 \cdot 50 - 7 \cdot 8 = 2,450 - 56 = 2,394 = 2,400 - 6$$

$$73(2,400 - 6) = 175,200 - 438 = 174,762$$

➢ Stretch Out Method

If two of the three numbers multiply to something that has the opposite last digit as the third number, multiply those two numbers first. For example, if you're multiplying $43 \cdot 36 \cdot 78$, you can tell just by the last digit that $43 \cdot 78$ ends in 4. After figuring out that $43 \cdot 78 = 3,354$ using the Negative Stretch Out Method, you can then multiply by 36 using the Stretch Out Method with both estimates being divisible by 10.

$$43 \cdot 78 = 40 \cdot 75 + 3 \cdot 118 = 3,000 + 354 = 3,354$$

$$3,354 \cdot 36 = 3,350 \cdot 40 - 4 \cdot 3,314 = 134,000 - 13,256 = 120,744$$

➢ Negative Stretch Out Method

If two of the three numbers multiply to something that has the same last digit as the third number, multiply those two numbers first. For example, if you're multiplying $43 \cdot 36 \cdot 78$, you can tell just by the last digit that $43 \cdot 36$ ends in 8. After figuring out that $43 \cdot 36 = 1,548$ using the Stretch Out Method, you can then multiply by 78 using the Negative Stretch Out Method with both estimates being divisible by 10.

$$43 \cdot 36 = 40 \cdot 39 - 4 \cdot 3 = 1,560 - 12 = 1,548$$

$$1,548 \cdot 78 = 1,550 \cdot 80 - 2 \cdot 1,628 = 124,000 - 3,256 = 120,744$$

➢ Proportional Stretch Out Method

You can always multiply any two of the numbers then use the Proportional Stretch Out Method. It's easiest if the number you save for last is close to something that's easy to multiply by, like a multiple of 25. For example, if you're multiplying 38·47·59, you can first multiply the 38 by 59 using the Stretch Out Method, then multiply the result by 47 using the Proportional Stretch Out Method.

$38·59 = 47·50 - 9·12 = 2,350 - 108 = 2,242$

$47·2,242$; *ratio* 45:1; *proportional estimate* 50, 2,250;
stretch out estimate 50, 2,107

$47·2,242 = 50·2,107 + 3·8 = 105,350 + 24 = 105,374$

➢ **Factor Into Two Numbers Instead Of Three**

If you're better at multiplying two 3-digit numbers than multiplying three 2-digit numbers, you can split one of the numbers by factoring. For example, if you're multiplying 29·53·78, you can factor the 78 into 6·13, then multiply that by the other numbers.

$29·53·78 = (29·13)(53·6) = 377·318$

Sgl Est: $377·318 = 300^2 + 300(77 + 18) + 77·18$
$= 90,000 + 28,500 + 1,386 = 119,886$

Prop Str Out: (*Prop Est* 400, 300); $377·318 = 401·300 - 23·18$
$= 120,300 - 414 = 119,886$

CHAPTER

9

CHECKING YOUR WORK

Since it's normal to make mistakes when doing math, it's important to know how to check your work. One way to check your work on a multiplication problem is to see if your solution is divisible by one of the numbers you multiplied.

> ➤ **How Divisibility Rules Work**

- Except for divisibility by 2 and 5 (and any numbers only divisible by powers of 2 and 5), all of the digits must be used in the rule. If a rule mentions the sum of or difference between the ones place and a higher place, the higher place includes itself and everything above it, while the ones place includes everything below the higher place. For example, if the thousands place and the ones place are mentioned, the number 12,345 would have a thousands place value of 12 and a ones place value of 345. If the hundreds place and the ones place are mentioned, it would have a hundreds place value of 123 and a ones place value of 45.

- If a number fits the divisibility rules of several different numbers with no common factors, that number is divisible by the product of those numbers. For example, a number divisible by 2 and 3 is divisible by 6. A number divisible by 3, 5, and 7 is divisible by 105.

- Subtracting can be reversed, because a positive number being divisible by something means that the same number being negative would also be divisible. For example, if a rule says twice the hundreds minus the ones, you could also reverse it by doing the ones minus twice the hundreds.

- If you follow one of the rules but still can't tell whether it's divisible, you can follow the rule again or follow another rule with the result. For example, to test whether 12,488 is divisible by 7, you could follow the rule that "thousands place minus ones place divisible by 7", leaving 476. You can now test the number 476 by following another rule, like "hundreds place x2 plus ones place divisible by 7", leaving 84.

- To test whether a number is divisible by something, you can divide that number by any other number (with no common factors) and it would still be divisible as long as it's a whole number. For example, to test whether a number is divisible by 7, you can divide it by any number not divisible by 7 and the result would still be divisible by 7 if it's a whole number.

- If a number is divisible, you can add or subtract any divisible number and it will still be divisible. This means you can shrink the places until they're as small as possible before using the divisibility rules. For example, if you're checking whether 9,452 is divisible by 17 and you're using the "hundreds place x2 minus ones place" rule, it may be hard to do 94·2-52. However, you can subtract 85 (5·17) from the 94 and 51 (3·17) from the 52, leaving 901, and it's easier to do 9·2-1=17.

- 0 is divisible by anything. If a rule results in 0, the number is divisible.

- If you're testing the divisibility of a number that's slightly less than the higher place, you can round up the higher place, and the ones place can be negative. For example, if the rule mentions the hundreds place, 697 can be interpreted as [7][0][-3] (seven hundred and negative three).

➢ **Rule Table**

For each rule in the list, the numbers in parentheses represent the multiple that the rule came from. The rule works because the number in parentheses is a multiple. Knowing this table will allow you to create your own divisibility rules, which will be shown after the list of rules.

MULTIPLE	RULE
exactly 2 place values exist $a \cdot 10^x + b$	a rule that results in 0 using those place values
power of 10 10^x	only place values less than 10^x matter
all 9s $10^x - 1$	if you're testing a number with x digits, the last digit can be moved to the beginning or the first digit can be moved to the end
1s only on the ends, 0s in the middle $10^x + 1$	same as for 10^x-1, but you must invert (make negative) either every digit you move or every digit you don't move
all 1s $\dfrac{10^x - 1}{9}$	subtract any amount from x consecutive digits

➤ **Divisibility Rule List For Numbers 2-37**

- Powers of 2
    - 2: Last digit divisible by 2. (10)
    - 4: Last two digits divisible by 4. (100)
    - 8: Last three digits divisible by 8. (1,000)
    - 16: Last four digits divisible by 16. (10,000)
    - etc
- Powers of 3
    - 3: Sum of digits divisible by 3. (9, 99, 999, 9,999, etc)
    - 9: Sum of digits divisible by 9. (9, 99, 999, 9,999, etc)
    - Pattern does not work with higher powers of 3.
- Powers of 5
    - 5: Last digit divisible by 5. (10)
    - 25: Last two digits divisible by 25. (100)

- 125: Last three digits divisible by 125. (1,000)
- 625: Last four digits divisible by 625. (10,000)
- etc

- 7
  - Hundreds place x2 plus ones place divisible by 7. (98)
  - Hundreds place x5 minus ones place divisible by 7. (105)
  - Hundreds place minus 3x ones place divisible by 7. (301)
  - Hundreds place plus 4x ones place divisible by 7. (399)
  - Thousands place minus ones place divisible by 7. (1,001)
  - If three digits, last digit moved to beginning and other digits inverted (made negative), or first digit moved to end and inverted divisible by 7. For example, if 616 is divisible by 7, so are [6][-6][-1]=539 and [1][6][-6]=154. (1,001)

- 11
  - Sum of odd places (ones, hundreds, ten thousands, etc) minus sum of even places (tens, thousands, hundred thousands, etc) divisible by 11. (11, 99, 1,001, 9,999, 100,001, etc)

- 13
  - Hundreds place x4 minus ones place divisible by 13. (104)
  - Hundreds place plus 3x ones place divisible by 13. (299)
  - Thousands place minus ones place divisible by 13. (1,001)
  - If three digits, last digit moved to beginning and other digits inverted, or first digit moved to end and inverted divisible by 13. For example, if 416 is divisible by 13, so are [6][-4][-1]=559 and [1][6][-4]=156. (1,001)
  -

- 17
  - Hundreds place x2 minus ones place divisible by 17. (102)
  - Thousands place x3 minus ones place divisible by 17. (1,003)
- 19
  - Hundreds place plus 4x ones place divisible by 19. (399)
  - Hundreds place x5 plus ones place divisible by 19. (95)
  - Thousands place x7 minus ones place divisible by 19. (1,007)
- 23
  - Hundreds place plus 3x ones place divisible by 23. (299)

- Hundreds place x8 plus ones place divisible by 23. (92)
- Thousands place x11 plus ones place divisible by 23. (989)
- Thousands place x12 minus ones place divisible by 23. (1,012)

- 27
  - Thousands place plus ones place divisible by 27. (999)
  - If three digits, last digit moved to beginning or first digit moved to end divisible by 27. For example, if 621 is divisible by 27, so are 216 and 162. (999)
  - Number divided by 3 has sum of digits divisible by 9.

- 29
  - Thousands place minus 2x ones place divisible by 29. (2,001)
  - Hundreds place x3 minus 2x ones place divisible by 29. (203)

- 31
  - Hundreds place x7 plus ones place divisible by 31. (93)
  - Thousands place x8 plus ones place divisible by 31. (992)

- 37
  - Thousands place plus ones place divisible by 37. (999)
  - If three digits, last digit moved to beginning or first digit moved to end divisible by 37. For example, if 925 is divisible by 37, so are 259 and 592. (999)
  - Any three consecutive digits reduced by the same amount is divisible by 37. For example, if 4,292 is divisible by 37, so are 2,072 (4,292-2,220) and 4,070 (4,292-222). It's easiest to pick the smallest digit to subtract so one of the digits becomes 0 and no carrying is necessary. For example, for the number 962, subtract 222 to get 740. For the number 814, subtract 111 to get 703. (111)

> **Creating Your Own Divisibility Rules**

If you forget any of those rules, you can create your own rule from a known multiple. To create a rule to test divisibility by n:

- you must compare exactly two place values; the rule is "something times one place value plus something else times the ones place is divisible by n"

- n must be prime and not 2 or 5
- neither of the two place values can be divisible by n
- your rule must not multiply either of the place values by something divisible by n

For example, since 98 (or 100-2) is divisible by 7, any rule resulting in a multiple of 7 by using 1 as the hundreds place and -2 as the ones place is accurate, as long as you're not testing a place value divisible by 7 or testing 7 times a place value. Since 1x ones plus 2x hundreds, 4x ones plus 1x hundreds, and 3x ones minus 1x hundreds all result in a multiple of 7 when testing a hundreds place of 1 and ones place of -2, those rules will also work when testing any multiple of 7.

➤ **Creating Your Own Divisibility Rules** (equation form)

Variable Meanings:

$(a \cdot 10^x + b)$ *is a known multiple of n*
"*p times 10^x place plus q times ones place divisible by n*" *is your rule*
$(c \cdot 10^x + d)$ *is a number you're testing with your rule*

If:

$(a \cdot 10^x + b)$ *divisible by n*
$(a \cdot p + b \cdot q)$ *divisible by n*
$(c \cdot p + d \cdot q)$ *divisible by n*
a, b, c, d, p, q all not divisible by n

Then:

$(c \cdot 10^x + d)$ *divisible by n*

➤ **Modulo Test**

Another option is to do the Modulo Test. Pick a modulo number and do the modulo operation for each of the numbers in the equation. If you did the original problem correctly, the modulo test will work no matter what modulo number you picked. If the multiplication of the remainders results in something greater than the modulo number, do the modulo operation of the result.

The easiest modulo number is 10, because modulo 10 is just the last digit. For example, if you did 376·823 and got 309,448, the modulo test results in 6·3=8, which passes because 6·3=18 and 18 modulo 10=8.

The next easiest modulo numbers are 9 and 11. For modulo 9, the remainder is the sum of the digits. Repeat if you get something greater than 9. For modulo 11, the remainder is the sum of all the odd placed digits (ones, hundreds, ten thousands, millions, etc) minus the sum of all the even placed digits (tens, thousands, hundred thousands, etc). Repeat if you get something greater than 11. Add 11 if negative.

In the example given before (376·823=309,448), modulo 9 results in 7·4=1 (3+7+6=16, 1+6=7; 8+2+3=13, 1+3=4; 3+0+9+4+4+8=28, 2+8=10, 1+0=1), which passes because 7·4=28 and 28 modulo 9=1. Modulo 11 results in 2·9=7 (3-7+6=2; 8-2+3=9; -3+0-9+4-4+8=-4, -4+11=7), which passes because 2·9=18 and 18 modulo 11=7.

> ## ➢ Alteration Test

If you subtract something from each of the numbers you multiplied (does not have to be the same amount subtracted from each number) and the result is all of the numbers being divisible by something, the product of the numbers subtracted can be subtracted from the original product and that result will also be divisible. In equation form:

Assume you multiplied x by y to get xy.

If (x – a) and (y – b) are both divisible by n, then
(xy – ab) is also divisible by n.

For example, assume that you were doing 237·584 and got 138,408. If you subtract 3 from 237 to get 234 and if you subtract 12 from 584 to get 572, both are divisible by 13, so you can subtract 36 (3·12) from 138,408 to get 138,372 and that should also be divisible by 13, which it is. If it's not, you'd have proven that 237·584 does not equal 138,408. It's usually good to test more than one modulo number, because a wrong answer can still sometimes pass the test (you could be off by an amount divisible by the modulo number), even though a correct answer will always pass the test. If you subtract 6 from 237 to get 231 and if you subtract 3 from 584 to get 581, both are divisible by 7, so you can subtract 18 (6·3) from 138,408 to get 138,390 and that should also be divisible by 7, which it is. If it's not, you'd have proven that 237·584 does not equal 138,408.

This rule works even if you're multiplying more than two numbers. For example, if you're multiplying 23·26·27, you can subtract 2, 5, and 6 to make each divisible by 7, then subtract 60 (2·5·6) from the product and test if that's also divisible by 7. The product is 16,146, and 16,086 (the product minus 60) is divisible by 7.

CHAPTER

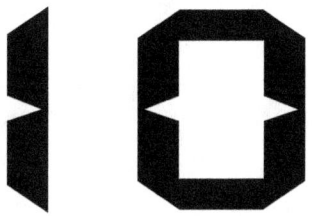

DIVISION AND DECIMALS

This chapter involves dividing by numbers that make repeating decimals. After reading this chapter, you'll be able to divide by any number up to 50, accurate to a large number of decimal places. Some numbers may be difficult because they have a large number of repeating digits (43 and 47 are the most difficult), so you can skip those if you can't memorize that many digits.

➢ **Basic Decimal Rules**

Fractions with a denominator with only prime factors of 2 and 5 are terminating decimals, with the same number of digits after the decimal point as the exponent of the 2 or the 5, whichever is greater.

$$\frac{1}{25} = \frac{1}{5^2} = 0.04 \ (2 \ digits, because \ 5^2)$$

$$\frac{1}{80} = \frac{1}{2^4 \cdot 5} = 0.0125 \ (4 \ digits, because \ 2^4)$$

Fractions with a denominator with prime factors other than 2 and 5 are repeating decimals. If there are no prime factors of 2 and 5, the repeating starts immediately after the decimal point. A bar over a set of numbers means that the sequence repeats forever.

$$\frac{1}{3} = 0.\overline{3}$$

$$\frac{1}{11} = 0.\overline{09}$$

$$\frac{1}{39} = 0.0\overline{25641}$$

Fractions with a denominator with prime factors of 2 and 5 and also other prime factors start the repeating late, with some digits not repeating. The number of non-repeating digits is equal to the exponent of the 2 or the 5, whichever is greater. In other words, the number of non-repeating digits is the same as how many there would be if the 2s and 5s were the only factors.

$$\frac{1}{60} = \frac{1}{2^2 \cdot 3 \cdot 5} = 0.01\overline{6} \ (2 \ non-repeating \ digits \ because \ 2^2)$$

$$\frac{1}{875} = \frac{1}{5^3 \cdot 7} = 0.001\overline{142857} \ (3 \ non-repeating \ digits \ because \ 5^3)$$

➢ **Converting Decimals To Fractions**

To convert a decimal into a fraction, the denominator is 10 to the power of the number of total digits, minus 10 to the power of the number of non-repeating digits. You can also think of the denominator as a 9 for every repeating digit and a 0 for every non-repeating digit. The numerator is the number after the decimal point as it appears, minus the non-repeating part. This method only works if there is at least one repeating digit, because this method will result in 0/0 if you use it for a terminating decimal.

$$0.1\overline{724} = \frac{1,724 - 17}{10,000 - 100} = \frac{1,707}{9,900}$$

$$0.3\overline{4298} = \frac{34{,}298 - 3}{100{,}000 - 10} = \frac{34{,}295}{99{,}990}$$

$$0.\overline{276} = \frac{276 - 0}{1{,}000 - 1} = \frac{276}{999}$$

➤ **Adding And Subtracting Decimals**

In order to add or subtract decimals, they have to have the same number of repeating and terminating digits. If they don't, you can rewrite the numbers so they do.

If a number terminates, it can have any arbitrary number of repeating zeros.

$$0.17 = 0.17\overline{0} = 0.17\overline{00} = 0.17\overline{000}$$

To increase the number of terminating digits, make the first repeating digit terminating, then place that digit at the end of the repeating sequence.

$$0.1\overline{234} = 0.12\overline{341} = 0.123\overline{412}$$

To increase the number of repeating digits, you can repeat the entire sequence.

$$0.\overline{1234} = 0.1234\overline{1234} = 0.1234\overline{12341234}$$

If the leftmost repeating digit does not need to carry, you can add or subtract the same way you would with a terminating decimal.

$$0.\overline{91} - 0.\overline{45} = 0.\overline{46}$$

$$0.\overline{34} + 0.\overline{42} = 0.\overline{76}$$

If the leftmost repeating digit needs to carry, you'll have to carry from the rightmost terminating digit, and do the same to the rightmost repeating digit. You have to carry from the rightmost repeating digit because every repetition will need to carry. In the following example, the terminating part adds to 5 and the repeating part adds to 133. The leading 1 in 133 has to carry, so you'll add 1 to the rightmost terminating digit and to the rightmost repeating digit, making the terminating part 6 and the repeating part 34.

$$0.2\overline{47} + 0.3\overline{86} = 0.6\overline{34}$$

> **Reciprocals**

For some numbers, like all 1s (111, 1111, 11111), or starting and ending with 1 (101, 1001, 10001), it's easy to find the reciprocal by finding a multiple that's all 9s, because the number 1 is equal to 0.999999999999999....

$$11 = \frac{99}{9}; \quad \frac{1}{11} = \frac{0.\overline{99}}{11} = 0.\overline{09}$$

$$101 = \frac{9,999}{99}; \quad \frac{1}{101} = \frac{0.\overline{9999}}{101} = 0.\overline{0099}$$

If your denominator has 3s instead of 1s, your decimal will have 3s instead of 9s. If your denominator has 9s instead of 1s, your decimal will have 1s instead of 9s. For example:

$$33 = \frac{99}{3}; \quad \frac{1}{33} = \frac{0.\overline{99}}{33} = 0.\overline{03}$$

$$909 = \frac{9,999}{11}; \quad \frac{1}{909} = \frac{0.\overline{9999}}{909} = 0.\overline{0011}$$

> **Cyclic Reciprocals**

If a number's reciprocal has n-1 repeating digits (for example, 1/7 has 6 repeating digits), every number from 1/n to (n-1)/n will have the same repeating digits in the same order, but starting from a different spot. Any number that follows this pattern is a cyclic number.

The sequences will always follow some kind of pattern, so if you can't memorize the sequences, strategies will be shown in the next section.

The first cyclic prime is 7. There are 6 (or 7-1) repeating digits.

$$\frac{1}{7} = 0.\overline{142857}$$

$$\frac{2}{7} = 0.\overline{285714}$$

$$\frac{3}{7} = 0.\overline{428571}$$

$$\frac{4}{7} = 0.\overline{571428}$$

$$\frac{5}{7} = 0.\overline{714285}$$

$$\frac{6}{7} = 0.\overline{857142}$$

This pattern also works with

$$\frac{1}{17} = 0.\overline{0588235294117647} \ (6\% \ upper \ estimate)$$

$$\frac{1}{19} = 0.\overline{052631578947368421} \ (5\% \ lower \ estimate)$$

$$\frac{1}{23} = 0.\overline{0434782608695652173913} \left(4\frac{1}{2}\% \ upper \ estimate, \ 4\frac{1}{3}\% \ lower \ estimate\right)$$

$$\frac{1}{29} = 0.\overline{0344827586206896551724137931}$$
$$\left(3\frac{1}{2}\% \ upper \ estimate, \ 3\frac{1}{3}\% \ lower \ estimate\right)$$

The estimates can help you figure out where to start the sequence. For example:

$$\frac{7}{17} \approx 6\%{\cdot}7 = 42\% \ upper \ estimate$$

closest to but below 42% is $0.\overline{4117647058823529}$

$$\frac{12}{19} \approx 5\%{\cdot}12 = 60\% \ lower \ estimate$$

closest to but above 60% is $0.\overline{631578947368421052}$

$$\frac{17}{23} \approx 4.5\%{\cdot}17 = 76.5\% \ upper \ estimate$$

closest to but below 76.5% is $0.\overline{7391304347826086956521}$

If you memorize those sequences of numbers, you can easily divide a number by 7, 17, 19, 23, or 29.

For 13, you'll have to memorize two different sets of numbers, and it's a 7.5% lower estimate or an 8% upper estimate.

$$\frac{1}{13} = 0.\overline{076923} \ (use \ for \ 1, 3, 4, 9, 10, 12)$$

$$\frac{2}{13} = 0.\overline{153846} \ (use \ for \ 2, 5, 6, 7, 8, 11)$$

For 31, you'll have to memorize two different sets of numbers, and it's a 3.25% upper estimate. The digits in one sequence are all 9 minus the digits in the other.

$$\frac{1}{31} = 0.\overline{032258064516129}$$

$$\frac{30}{31} = 0.\overline{967741935483870}$$

For 43, you'll have to memorize two different sets of numbers, and it's a 2 1/3 % upper estimate. The digits in one sequence are all 9 minus the digits in the other.

$$\frac{1}{43} = 0.\overline{023255813953488372093}$$

$$\frac{42}{43} = 0.\overline{976744186046511627906}$$

For 47, you'll have to memorize a single 46-digit sequence. It's a 2 1/8 % lower estimate.

$$\frac{1}{47} = 0.\overline{0212765957446808510638297872340425531914893617}$$

The number 49 uses a 42-digit sequence. It is almost a cyclic sequence, except for denominators that are multiples of 7, because those can be simplified. It's a 2% lower estimate.

$$\frac{1}{49} = 0.\overline{020408163265306122448979591836734693877551}$$

➤ **Cyclic Reciprocals: Memorization Strategies**

It may seem difficult to memorize all those sequences, but it's easier if you know that they follow patterns and are not just arbitrary numbers. If you notice something in the sequence that looks like a pattern, it probably is.

If a prime number's reciprocal has an even number of repeating digits, the last half of the digits will always be 9 minus the first half of the digits. Because of this rule, you really only have to memorize the first half of the digits.

7 and 13 are small sequences, so you can most likely memorize the entire sequence with no other tricks.

For 17, the first half is 05882352. Start with 05, and double and add 1 for the next two digits, and do the same two more times, which would give you 05112347. Then, take 9 minus each digit of the 11 and 47, which would give you 05882352.

For 19, the first half is 052631578. Start with 0526. The next two digits are 05+26=31. The next two are 26+31=57. Then, just remember the 8, and you have the first half of the digits.

For 23, the first half is 04347826086. The 086 is easy to remember because it's just double the first three digits, so you'll have to find some way of memorizing 04347826. You have several choices.

- Start with 26 on the right. The next two digits to the left are 26·3=78. The next two are 78·3=234, or 34 with 2 carrying into the next set. The next two are 34·3=102, then adding the carried 2 and leaving out the extra digit makes 04.
- Start with 0434 on the left. The next two digits are 2(04+34)+2=78 (the +2 is because the next set carries). The next two digits are

2(34+78)+2=226, or 26 with the 2 carrying into the previous set (the +2 is because the next set carries).

- Start with 043 on the left. To get the next three digits, divide those three digits by 2 (adding 5 to the first digit if the previous digit is odd), then take 9 minus each digit. Dividing by 2 gives you 021, plus 5 to the first digit because the previous digit is 3 (the previous digit is the end of the sequence, and you know it's 3 because the sequence will be all 9s when multiplied by 23), gives you 521, and 9 minus each digit is 478. Dividing that by 2 gives you 239, plus 5 to the first digit because the previous digit is 3 gives you 739, and 9 minus each digit is 260. You now have "043478260".

For 29, the first half is 03448275862068. The 068 is easy to remember because it's just double the first three digits, so you'll have to find some way of memorizing 03448275862. You have several choices.

- Start with 0344. The next two digits are 2(44-03)=82. The next two are 2(82-44)-1=75 (the -1 is because the next set carries). The next two are 2(75-82)+100=86. Then, just remember the 2 and double the first three digits, and you have the first half of the digits.
- Start with 034. To get the next three digits, divide those three digits by 2 (adding 5 to the first digit if the previous digit is odd), then take 9 minus each digit. Dividing by 2 gives you 017, plus 5 to the first digit because the previous digit is 1 (the previous digit is the end of the sequence, and you know it's 1 because the sequence will be all 9s when multiplied by 29), gives you 517, and 9 minus each digit is 482. Dividing that by 2 gives you 241, do not add 5 to the first digit because the previous digit is 4, and 9 minus each digit is 758. Dividing that by 2 gives you 379, do not add 5 to the first digit because the previous digit is 2, and 9 minus each digit is 620. You now have "034482758620".

For 31, one of the sequences is 032258064516129. The other sequence is just 9 minus those digits. Start with 032258. The next six digits are double the first six, which is 064516. The next six are double those, which is 129032, but the sequence already resets after three of those digits.

For 43, one of the sequences is 023255813953488372093. The other sequence is just 9 minus those digits. Start with the 093 on the right. The next three digits to the left are 093·4=372. The next three digits are 372·4=1,488, or 488 with the 1 carrying into the next set. The next three digits are 488·4+1=1,953, or 953 with 1 carrying into the next set. Continue the pattern.

For 47, the first half is 0212765957446808510382. Start with 0212765. The next seven digits are 9 minus the digits of double the first seven digits. Doubling results in 0425531 (0425530 plus 1 because the digit after the 5 is a 9, so you carry a 1), then 9 minus each digit results in 9574468. The next seven digits are 4 times the first seven digits (0851063, or 0851060 plus 3 carried). You'll also notice that 063 is triple the first three digits. Then, just remember the 82, and you have the first half of the digits.

For 49, the first half is 020408163265306122448. Every two digits is double the previous two, carrying whenever necessary.

> **Dividing By Parts** (cyclic and terminating)

To divide by one of the numbers given before (7, 11, 13, 17, 19, 23, 29, 31, 43, 47, 49) times a number that's only divisible by 2s and 5s, separate the fraction into two parts, one repeating and one terminating. The terminating part will have a denominator with prime factors of only 2s and 5s. The repeating part will have a denominator with other prime factors but not 2s and 5s. For example, if you're dividing by 56, separate it into 7 and 8. If you're dividing by 85, separate it into 17 and 5. If you're dividing by 1150, separate it into 23 and 50. The number that's only 2s and 5s will be known as the "terminating number" and the other number will be known as the "cyclic number".

For example, 11/35

$$\frac{11}{35} = \frac{}{7} + \frac{}{5}$$

To solve this, keep subtracting the terminating number until the numerator is divisible by the cyclic number. In this example, if you keep subtracting 5 from 11, you'll eventually get -14. You subtracted a total of 25 and ended with -14, so 11/35=25/35-14/35.

$$\frac{11}{35} = \frac{25}{35} - \frac{14}{35} = \frac{5}{7} - \frac{2}{5} = 0.\overline{714285} - 0.4 = 0.3\overline{142857}$$

Note that the terminating part only affects the beginning of the decimal, not every repetition. When this happens, the repetition starts late, so any digits that are changed at the beginning of the repetition must be moved to the end.

For example, 31/68

$$\frac{31}{68} = \frac{}{17} + \frac{}{4}$$

If you keep subtracting 4 from 31 until it's divisible by 17, you'll get -17 after 12 times of subtracting 4 for a total of 48.

$$\frac{31}{68} = \frac{48}{68} - \frac{17}{68} = \frac{12}{17} - \frac{1}{4} = 0.\overline{7058823529411764} - 0.25$$

$$= 0.45\overline{5882352941176470}$$

You could instead change the fraction so the terminating part is a power of 10. In the examples given, 11/35 could become 22/70, and 31/68 could become 775/1,700. To use this method, one denominator will be a power of 10, and one denominator will be the original fraction's denominator. You'll need to separate the original fraction into two parts, one of which can be simplified into a fraction with a denominator that's a power of 10.

$$\frac{22}{70} = \frac{21}{70} + \frac{1}{70} = \frac{3}{10} + \frac{1}{70} = 0.3 + 0.0\overline{142857} = 0.3\overline{142857}$$

$$\frac{775}{1,700} = \frac{765}{1,700} + \frac{10}{1,700} = \frac{45}{100} + \frac{10}{1,700} = 0.45 + 0.005\overline{882352941176470}$$

$$= 0.45\overline{5882352941176470}$$

> ➢ **Dividing By Parts** (cyclic and single-digit repeating)

If you're dividing by a cyclic prime times 3, 9, or 11, you can do a similar method as done before. You'll either be adding something to every repeating digit (if 3 or 9) or alternating between adding and subtracting (if 11). It's hard to do this with long sequences, so it's usually only possible to do in your head if the cyclic prime is either 7 or 13, since both have only six repeating digits.

For example, 41/63

$$\frac{41}{63} = \frac{}{7} + \frac{}{9}$$

After subtracting 7 twice for a total of 14, you get 27, which is 3·9.

$$\frac{41}{63} = \frac{14}{63} + \frac{27}{63} = \frac{3}{7} + \frac{2}{9} = 0.\overline{428571} + 0.\overline{2}$$

To solve this, add 2 to each digit of 428571, carrying when needed.

$$\frac{41}{63} = \frac{3}{7} + \frac{2}{9} = 0.\overline{428571} + 0.\overline{2} = 0.\overline{428571} + 0.\overline{222222} = 0.\overline{650793}$$

For example, 79/143

$$\frac{79}{143} = \frac{}{13} + \frac{}{11}$$

After subtracting 11 six times for a total of 66, you get 13, which is 1·13.

$$\frac{79}{143} = \frac{66}{143} + \frac{13}{143} = \frac{6}{13} + \frac{1}{11} = 0.\overline{461538} + 0.\overline{09}$$

To solve this, add 1 to the first digit and alternate between adding 1 and subtracting 1 to the digits of 461538.

$$\frac{79}{143} = \frac{6}{13} + \frac{1}{11} = 0.\overline{461538} + 0.\overline{09} = 0.\overline{461538} + 0.\overline{090909} = 0.\overline{552447}$$

Unlike when dividing by 2 or 5, both parts are fully repeating, so the repetition is not delayed.

> ➢ **Special Case: Dividing By 27 and 37**

Since 27 and 37 are both factors of 999, you don't have to memorize a long sequence to divide by 27 or 37. You can just multiply by $0.\overline{037}$ or $0.\overline{027}$.

$$\frac{17}{27} = 17 \cdot 0.\overline{037} = 0.\overline{629}$$

$$\frac{13}{37} = 13 \cdot 0.\overline{027} = 0.\overline{351}$$

➢ **Special Case: Dividing By 41**

Unlike most of the prime numbers shown in this chapter, 41 is a rare example that only has five repeating digits. $1/41 = 0.\overline{02439}$. You could try memorizing 8 different cyclic sequences (02439, 04878, 07317, 09756, 12195, 14634, 26829, and 36585), but it would most likely be easier to just multiply by 2,439.

To multiply a number by 2,439, you can multiply it by 271 then by 9 (see Chapter 4 for multiplying by 9). To multiply by 271, you can multiply by 270 and add your original number.

$$\frac{29}{41} = 29 \cdot 0.\overline{02439} = 29 \cdot 0.\overline{00271} \cdot 9$$

$$29 \cdot 0.\overline{00271} \cdot 9 = 29(0.\overline{00270} + 0.\overline{00001}) \cdot 9 = 0.\overline{07859} \cdot 9 = 0.\overline{70731}$$

You can make it easier by first taking the opposite of the number (1 minus the fraction) if the numerator is 21 through 40, then take the opposite again (1 minus the result) at the end. This will make it so you never have to multiply 271 by anything greater than 20.

$$\frac{29}{41} = 1 - \frac{12}{41}$$

$$\frac{12}{41} = 12 \cdot 0.\overline{00271} \cdot 9 = 0.\overline{03252} \cdot 9 = 0.\overline{29268}$$

$$\frac{29}{41} = 1 - \frac{12}{41} = 0.\overline{99999} - 0.\overline{29268} = 0.\overline{70731}$$

➢ **Improper Fractions**

If your original fraction is an improper fraction, turn it into a mixed number first. For example:

$$\frac{139}{7} = 19\frac{6}{7}$$

$$\frac{6}{7} = 0.\overline{857142}$$

$$\frac{139}{7} = 19.\overline{857142}$$

> ➤ **Review**

Using the methods shown in this chapter, you can now divide by any number up to 50.

Number	How To Divide
2	Multiply by 0.5
3	Multiply by $0.\overline{3}$
4	Multiply by 0.25
5	Multiply by 0.2
6	Divide by parts (2 and 3)
7	Use cyclic sequence 142857
8	Multiply by 0.125
9	Multiply by $0.\overline{1}$
10	Multiply by 0.1
11	Multiply by $0.\overline{09}$
12	Divide by parts (3 and 4)
13	Use cyclic sequences 076923, 153846
14	Divide by parts (2 and 7)
15	Divide by parts (3 and 5)
16	Multiply by 0.0625
17	Use cyclic sequence 0588235294117647
18	Divide by parts (2 and 9)
19	Use cyclic sequence

		052631578947368421
20		Multiply by 0.05
21		Divide by parts (3 and 7) Or use cyclic sequences 047619, 095238, 142857
22		Divide by parts (2 and 11)
23		Use cyclic sequence 0434782608695652173913
24		Divide by parts (3 and 8)
25		Multiply by 0.04
26		Divide by parts (2 and 13)
27		Multiply by $0.\overline{037}$
28		Divide by parts (4 and 7)
29		Use cyclic sequence 0344827586206896551724137931
30		Divide by parts (3 and 10)
31		Use cyclic sequences 032258064516129, 967741935483870
32		Multiply by 0.03125
33		Multiply by $0.\overline{03}$
34		Divide by parts (2 and 17)
35		Divide by parts (5 and 7)
36		Divide by parts (4 and 9)
37		Multiply by $0.\overline{027}$
38		Divide by parts (2 and 19)
39		Divide by parts (3 and 13) Or use cyclic sequences 025641, 051282, 076923
40		Multiply by 0.025
41		Multiply by $0.\overline{02439}$
42		Divide by parts (2, 3, and 7; or 2 and 21)
43		Use cyclic sequences 023255813953488372093, 976744186046511627906
44		Divide by parts (4 and 11)

45	Divide by parts (5 and 9)
46	Divide by parts (2 and 23)
47	Use cyclic sequence 0212765957446808510638297872340425531914893617
48	Divide by parts (3 and 16)
49	Simplify if numerator divisible by 7 If not, use sequence 020408163265306122448979591836734693877551
50	Multiply by 0.02

CHAPTER

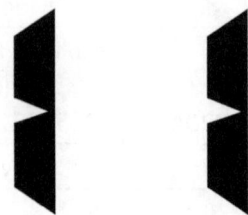

SPECIAL CASE MULTIPLICATION

In previous chapters, you learned general methods for multiplying arbitrary numbers. In this chapter, you'll learn about special cases involving the multiplication of a specific number. All will use the Almost-Round Multiple Method, with the steps adjusted for the specific numbers.

➢ **Multiplying by 143 and 167**

Multiplying by 143 is the same as dividing by 7 then multiplying by 1,001. Multiplying by 167 is the same as dividing by 6 then multiplying by 1,002. There are three methods for multiplying a 3-digit number by 143 or 167. The first method is to change the other number into something times 6 (for 167) or 7 (for 143) plus a remainder. You'll end up with something times 1,001 or 1,002, plus something times 143 or 167.

$$143 \cdot 583 = 143(7 \cdot 83 + 2) = 1,001 \cdot 83 + 143 \cdot 2$$
$$= 83,083 + 286 = 83,369$$

$$167 \cdot 583 = 167(6 \cdot 97 + 1) = 1,002 \cdot 97 + 167 \cdot 1$$

$$= 97{,}194 + 167 = 97{,}361$$

The second method is to first separate the hundreds place, then do the same as before with just the tens and ones place.

$$143 \cdot 583 = 143(500 + 7 \cdot 11 + 6) = 71{,}500 + 11{,}011 + 858 = 83{,}369$$

$$167 \cdot 583 = 167(500 + 6 \cdot 13 + 5) = 83{,}500 + 13{,}026 + 835 = 97{,}361$$

Normally, it's easier to do the first method if the number is divisible (by 6 if multiplying by 167, by 7 if 143), and the second method if it's not divisible.

For 143, it's good to memorize the multiples 143, 286, 429, 572, 715, and 858 (all are 1 plus any three consecutive numbers in the sequence 1-4-2-8-5-7). For 167, it's good to memorize the multiples 167, 334, 501, 668, and 835.

The third method is to multiply by 1,001 or 1,002 first, then divide by 7 or 6.

$$143 \cdot 583 = \frac{1{,}001}{7} \cdot 583 = \frac{583{,}583}{7} = 83{,}369$$

$$167 \cdot 583 = \frac{1{,}002}{6} \cdot 583 = \frac{584{,}166}{6} = 97{,}361$$

➢ **Cyclic Number Multiplication: 142,857**

Due to the nature of repeating decimals, cyclic numbers can be easier to multiply than other numbers similar in size.

142,857 is the repeating pattern when a whole number is divided by 7, and $142{,}857 \cdot 7 = 999{,}999$. Therefore, you can multiply a number by 142,857 by dividing by 7 then multiplying by 999,999 (1,000,000-1).

> **Step 1:** Divide by 7 and keep track of the whole part and the remainder.
> **Step 2:** Multiply the remainder by 142,857. The multiples of 142,857 all have the same digits in the same order, but they start in different places (142,857, 285,714, 428,571, 571,428, 714,285, 857,142).
> **Step 3:** Add 1,000,000 times the whole part to your result from Step 2 and subtract 1 times the whole part. The 1,000,000s place will always start immediately before the six digits from Step 2.

586·142,857

Step 1: $\dfrac{586}{7} = 83\dfrac{5}{7}$

Step 2: 5·142,857 = 714,285

Step 3: 586·142,857 = 714,285 + 83,000,000 – 83 = 83,714,202

417·142,857

Step 1: $\dfrac{417}{7} = 59\dfrac{4}{7}$

Step 2: 4·142,857 = 571,428

Step 3: 417·142,857 = 571,428 + 59,000,000 – 59 = 59,571,369

665·142,857

Step 1: $\dfrac{665}{7} = 95$

Step 2: 0·142,857 = 000,000

Step 3: 665·142,857 = 95,000,000 – 95 = 94,999,905

➤ **Cyclic Number Multiplication: Multiplying By 588,235,294,117,647**

This trick works with any cyclic number (see Chapter 10). Examples include 52,631,578,947,368,421 (999,999,999,999,999,999/19) and 434,782,608,695,652,173,913 (9,999,999,999,999,999,999,999/23).

Step 1: Divide by 17 (or whatever number is needed to multiply to result in all 9s if you're using a different cyclic number) and keep track of the whole part and the remainder.

Step 2: Multiply the remainder by 588,235,294,117,647 (or whatever the cyclic number is). The multiples of 588,235,294,117,647 all have the same digits in the same order, but they start in different places (588,235,294,117,647, 1,176,470,588,235,294, 1,764,705,882,352,941, 2,352,941,176,470,588, etc). The cyclic number actually has a leading 0 (pretend it's 0,588,235,294,117,647, 052,631,578,947,368,421, or 0,434,782,608,695,652,173,913).

Step 3: Add 10^{16} (10^{18} for 19, 10^{22} for 23) times the whole part to your result from Step 2 and subtract 1 times the whole part. The 10^{16} (or 10^{18} or 10^{22}) place will always start immediately before the digits from Step 2.

927·588,235,294,117,647

$Step\ 1: \dfrac{927}{17} = 54\dfrac{9}{17}$

$Step\ 2:$ 9·0,588,235,294,117,647 = 5,294,117,647,058,823

$Step\ 3:$ 927·588,235,294,117,647
= 5,294,117,647,058,823 + 540,000,000,000,000,000 – 54
= 545,294,117,647,058,769

749·588,235,294,117,647

$Step\ 1: \dfrac{749}{17} = 44\dfrac{1}{17}$

$Step\ 2:$ 1·0,588,235,294,117,647 = 0,588,235,294,117,647

$Step\ 3:$ 749·588,235,294,117,647
= 0,588,235,294,117,647 + 440,000,000,000,000,000 – 44
= 440,588,235,294,117,603

> ➢ **Cyclic Number Multiplication: Multiplying By 76,923**

76,923 is 999,999/13, which is not quite a cyclic number because only half of its multiples use the sequence 0-7-6-9-2-3. The other half use 1-5-3-8-4-6.

Step 1: Divide by 13 and keep track of the whole part and the remainder.
Step 2: Multiply the remainder by 76,923. The 1st, 3rd, 4th, 9th, 10th, and 12th multiples use the sequence 0-7-6-9-2-3, while the 2nd, 5th, 6th, 7th, 8th, and 11th multiples use the sequence 1-5-3-8-4-6.
Step 3: Add 1,000,000 times the whole part to your result from Step 2 and subtract 1 times the whole part. The 1,000,000s place will always start immediately before the six digits from Step 2.

737·76,923

$Step\ 1: \dfrac{737}{13} = 56\dfrac{9}{13}$

$Step\ 2: 9 \cdot 076{,}923 = 692{,}307$

$Step\ 3: 737 \cdot 76{,}923 = 692{,}307 + 56{,}000{,}000 - 56 = 56{,}692{,}251$

$658 \cdot 76{,}923$

$Step\ 1: \dfrac{658}{13} = 50\dfrac{8}{13}$

$Step\ 2: 8 \cdot 076{,}923 = 615{,}384$

$Step\ 3: 658 \cdot 76{,}923 = 615{,}384 + 50{,}000{,}000 - 50 = 50{,}615{,}334$

➢ **First-Half Cyclic Number Multiplication**

Multiplying by the first half of a cyclic number (with the last digit rounded up) is very similar to multiplying by the entire cyclic number. The critical differences are that you round up the last digit of the sequence in Step 2, and that you add 1 times the whole part instead of subtracting it in Step 3. The second half is always 9 minus every digit of the first half, and the first half stops immediately before that happens.

Cyclic Number	First Half
142,857	143
0,588,235,294,117,647	05,882,353
052,631,578,947,368,421	052,631,579
0,434,782,608,695,652,173,913	04,347,826,087
0,344,827,586,206,896,551,724,137,931	03,448,275,862,069

143 was already shown before, so 5,882,353 will be used in this example. The steps can be adjusted for other cyclic numbers.

Step 1: Divide by 17 and keep track of the whole part and the remainder.
Step 2: Multiply the remainder by 5,882,353. The multiples of 5,882,353 all follow the sequence 0-5-8-8-2-3-5-2-9-4-1-1-7-6-4-7, keeping only the first half (first 8 digits) and rounding the last digit up (11,764,706, 17,647,059, 23,529,412, 29,411,765, etc).

Step 3: Add 100,000,000 times the whole part to your result from Step 2 and also add 1 times the whole part. The 100,000,000s place will always start immediately before the eight digits from Step 2.

632·5,882,353

$Step\ 1: \dfrac{632}{17} = 37\dfrac{3}{17}$

$Step\ 2: 3{\cdot}05{,}882{,}353 = 17{,}647{,}059$

$Step\ 3: 632{\cdot}5{,}882{,}353 = 17{,}647{,}059 + 3{,}700{,}000{,}000 + 37 = 3{,}717{,}647{,}096$

CHAPTER

DAYS BETWEEN DATES

Have you ever wondered how to figure out how many days are between today and another date, or what day of the week another date was? This chapter will teach you that.

> ## Dates Less Than A Year Apart

To figure out the number of days between dates less than a year apart:

Step 1: Figure out the difference in months, and multiply by 31. If your ending date has an earlier month than your starting date, add 12 to the ending date's month and subtract 1 from the ending date's year.

Step 2: Subtract 1 for each time April, June, September, or November ends in your range. Subtract 2 for each leap-year February and 3 for each regular February that ends in your range. A month ends in your range if and only if your range includes both the last day of that month and the first day of the next month.

Step 3: Add the difference between the day numbers (sign matters, later date's day minus earlier date's day).

For example, from February 26, 2014 to November 12, 2014:

Step 1: $11 - 2 = 9$; $9 \cdot 31 = 279$

Step 2: *Subtract 3 for February and 1 for each of April, June, and September.*
$279 - 6 = 273$

Step 3: $273 + (12 - 26) = 259$

November 12, 2014 is 259 days after February 26, 2014.

> ➤ **Adding One Year**

To add an entire year, add 365 if the end of February in the year you're adding is not a leap year, 366 if it is a leap year. If you're adding multiple years, the steps will be shown after the following examples.

For example, from January 19, 2015 to April 3, 2016:

You can either do January 19, 2015 to April 3, 2015 plus a leap year, or January 19, 2016 to April 3, 2016 plus a regular year. The steps shown will be January 19, 2015 to April 3, 2015 plus a leap year.

Step 1: $4 - 1 = 3$; $31 \cdot 3 = 93$

Step 2: *Subtract 3 for February.* $93 - 3 = 90$

Step 3: $90 + (3 - 19) = 74$

Step 4: $74 + 366 = 440$

April 3, 2016 is 440 days after January 19, 2015.

> ➤ **Adding Multiple Years**

If you're adding multiple years, the easiest way to multiply by 365 is to use the almost-round multiple of 1,095 (or 1,100-5), which is $365 \cdot 3$.

Step 1: Change one of the years so your dates are less than a year apart (starting date must still be before ending date). Use the Days Between Dates Method to figure out the days between the dates.

Step 2: Divide the year difference (the number of years subtracted from your range in Step 1) by 3, keeping track of the whole part and the remainder.

Step 3: For the remainder from Step 2, multiply it by 365 and add it to the result from Step 1.

Step 4: For the whole part from Step 2, multiply it by 1,095 and add it to the result from Step 3. In other words, multiply the whole part by 1,100 and add, then multiply the whole part by 5 and subtract.

Step 5: Add one for each leap year. Your range for this step is between your original date and the adjusted date in Step 1. Do not count leap years that were already counted in Step 1.

To figure out the number of leap years, change one of the dates so the range is exactly divisible by 4 years, then divide by 4 years. If changing the date does not add or subtract a February 29 of a leap year, your adjusted range divided by 4 years is the number of leap years. If changing the date added a February 29 of a leap year, subtract 1. If changing the date subtracted a February 29 of a leap year, add 1.

For example, assume that today is February 21, 2015 and you were born on November 23, 1989, and you wanted to figure out how old you are in days. Since you're comparing an earlier date in a later year, treat February of 2015 as the 14th month of 2014.

Step 1a: $2 + 12 - 11 = 3$; $3 \cdot 31 = 93$

Step 1b: Subtract 1 for November. $93 - 1 = 92$

Step 1c: $92 + (21 - 23) = 90$

February 21, 2015 is 90 days after November 23, 2014.

Step 2: $\dfrac{25}{3} = 8\dfrac{1}{3}$

Step 3: $90 + 1 \cdot 365 = 90 + 365 = 455$

Step 4: $455 + 8{,}800 - 40 = 9{,}215$

Step 5a: Stretch to November 23, 1989 to November 23, 2013

Step 5b: No leap years in adjustment. $\dfrac{24 \text{ years}}{4 \text{ years}} = 6; 9{,}215 + 6 = 9{,}221$

February 21, 2015 is 9,221 days after November 23, 1989.

> ➢ **Day Of The Week**

To figure out what day of the week a certain date is, the steps are very similar to the steps used to figure out the number of days between dates, but only with the remainder when divided by 7.

> **Step 1:** Change one of the years so your dates are less than a year apart (starting date must still be before ending date).
> **Step 2:** Figure out the difference in months, and multiply by 3. If your ending date has an earlier month than your starting date, add 12 to the ending date's month and subtract 1 from the ending date's year.
> **Step 3:** Subtract 1 for each time April, June, September, or November ends in your range. Subtract 2 for each leap-year February and 3 for each regular February that ends in your range. A month ends in your range if and only if your range includes both the last day of that month and the first day of the next month.
> **Step 4:** Add the difference between the day numbers (sign matters, later date's day minus earlier date's day).
> **Step 5:** Add 1 for each year that you changed the date by in Step 1.
> **Step 6:** Add 1 for each leap year.
> **Step 7:** If your result is negative, keep adding 7 until your result is positive. Take the remainder of your result when divided by 7, and add that many days to the current day of the week if you're comparing today to a future date, subtract that many days from the current day of the week if you're comparing to a past date.

Using the previous examples:

If today is Wednesday, February 26, 2014 and you want to know what day November 12, 2014 is:

Step 2: $11 - 2 = 9; 9{\cdot}3 = 27$

Step 3: Subtract 3 for February and 1 for each of April, June, and September. $27 - 6 = 21$

Step 4: 21 + (12 − 26) = 7

Step 7: 7 *mod* 7 = 0; *Wednesday* + 0 = *Wednesday*

Steps 1, 5, and 6 were skipped because you're testing dates from the same year. November 12, 2014 is a Wednesday.

If today is Monday, January 19, 2015 and you want to know what day April 3, 2016 is:

Step 1: You can either do January 19, 2015 to April 3, 2015 plus a leap year, or January 19, 2016 to April 3, 2016 plus a regular year. The steps shown will be January 19, 2015 to April 3, 2015 plus a leap year.

Step 2: 4 − 1 = 3; 3·3 = 9

Step 3: *Subtract* 3 *for February*. 9 − 3 = 6

Step 4: 6 + (3 − 19) =− 10

Step 5: − 10 + 1 =− 9

Step 6: − 9 + 1 =− 8

Step 7: − 8 + 14 = 6; *Monday* + 6 = *Sunday*

April 3, 2016 is a Sunday.

> **Day Of The Week: End-Of-Century Reference Points**

The previously stated steps use today as the reference point, but you can use any date as the reference point as long as you know what day of the week it was. It may help to use Sunday, December 31, 1899 or Friday, December 31, 1999 as the reference point, because those are the end of the century. It's important to know that 1900 was not a leap year even though it's divisible by 4, because a year that is divisible by 100 must be divisible by 400 to be a leap year. 2000 was a leap year.

When figuring out how the month affects the day of the week, instead of using the usual Steps 2 and 3, you can just use this table. Subtract 1 for January and February of leap years.

Month	Effect On Weekday
January	0
February	3
March	3
April	6
May	1
June	4
July	6
August	2
September	5
October	0
November	3
December	5

The year's effect on the weekday would be the last 2 digits of the year, times 1.25, rounded down to the nearest integer. Add 1 if you're comparing to 2000, because 2000 is a leap year but 1900 is not.

The day's effect on the weekday would just be the day.

Add the month's effect, the year's effect, and the day's effect. Divide by 7 and keep the remainder. Add that many days to your reference point.

In the examples, "floor" means rounded down to the nearest integer.

June 20, 2015

- Reference Point: Friday, December 31, 1999
- Month's Effect: 4 (from the table)
- Day's Effect: 20
- Year's Effect: floor(15·1.25)+1=19
- Total Effect: 4+20+19=43; 43 mod 7=1
- Friday + 1 = Saturday; June 20, 2015 is a Saturday

April 14, 1957

- Reference Point: Sunday, December 31, 1899
- Month's Effect: 6 (from the table)
- Day's Effect: 14
- Year's Effect: floor(57·1.25)=71
- Total Effect: 6+14+71=91; 91 mod 7=0
- Sunday + 0 = Sunday; April 14, 1957 is a Sunday

> **Day Of The Week: End-Of-Century Reference Points**
> (Reference Point Later Than Target Date)

For years close to but before the end of a century, you can use the next century as the reference point and subtract the year's effect. Add 1 (subtract 1 less) if the year is divisible by 4. The day's and month's effect will remain the same.

November 23, 1989

- Reference Point: Friday, December 31, 1999
- Month's Effect: 3 (from the table)
- Day's Effect: 23
- Year's Effect: -floor(11·1.25)=-13
- Total Effect: 3+23-13=13; 13 mod 7=6
- Friday + 6 = Thursday; November 23, 1989 is a Thursday

> **Day Of The Week: 28 Years**

The number of days in 28 years is always exactly divisible by 7 (unless the 28 years includes the end of February in a year that's divisible by 100 but not by 400), so you can always add or subtract a multiple of 28 years without changing the day of the week (in the rare exception mentioned, the number of days would be one less than a multiple of 7). If you're testing the year 1961, you can instead subtract 56 and test the year 1905. If you're testing the year 2087, you can instead subtract 84 and test the year 2003. If you're testing the year 1974, you can instead add 28 and test the year 2002.

www.ingramcontent.com/pod-product-compliance
Lightning Source LLC
Chambersburg PA
CBHW070823180526
45168CB00002B/732